Communicating Science
a handbook

Communicating Science

a handbook

Michael Shortland
and
Jane Gregory

Copublished in the United States with
John Wiley & Sons, Inc., New York

Longman Scientific & Technical
Longman Group UK Limited
Longman House, Burnt Mill, Harlow
Essex CM20 2JE, England
and Associated Companies throughout the world.

Copublished in the United States with
John Wiley & Sons Inc., 605 Third Avenue, New York, NY 10158

First published in 1991

British Library Cataloguing-in-Publication Data
Shortland, Michael
Communicating science.
1. Communication
I. Title II. Gregory, Jane
302.2

ISBN 0-582-05709-4

Library of Congress Cataloging in Publication Data
Shortland, Michael.
Communicating science : a handbook / Michael Shortland and Jane Gregory.
p. cm.
Includes bibliographical references and index.
ISBN 0-470-21696-4
1. Communication in science. I. Gregory, Jane, 1962-
II. Title.
Q223.S48 1991
501.4--dc20

90-46395
CIP

Produced by Longman Group (FE) Limited
Printed in Hong Kong

Contents

Foreword

Modern society is science and technology based. Our industry, and so our national prosperity, are mostly dependent on science and technology. In the home and at work we use products of this industry and science affects nearly all policy issues of national and international importance. It also impinges on a wide range of personal activities, from health and diet to holidays and sport. The scientific revolution is continuing and promises to be of enormous benefit to us all. But if we are to achieve these potential benefits then we must have a scientifically literate, or as I have called it, scienceate population, that is ready to accept and exploit the fruits of science. Scientists themselves have a major responsibility in achieving this end. The Royal Society's Report of 1985 on the Public Understanding of Science, put this most succinctly in its final sentence "our most direct and urgent message must be to the scientists themselves: learn to communicate with the public, be willing to do so and consider it your duty to do so".

This handbook on communicating science makes a valuable contribution in providing a wide variety of practical advice to the scientist on how he or she should communicate with the public.

A non-scientific audience should be able to appreciate the elegance and excitement of science in just the same way that the artistry in a painting or the composition of a piece of music can be enjoyed. Science should be as much a part of our culture to enjoy as are the arts and humanities. Many surveys tell us that there is a public desire for more information about science and technology. Scientists must work with their colleagues in the media and elsewhere to improve and increase the information that is available, so that this need can be properly satisfied.

So many of the policies that are thrust upon us by the various political parties depend on science and technology, and many have major effects on our prosperity and direct effects on our everyday lives. Without some understanding of the science underlying these issues how, for example, can the public cast an informed vote on polling day for a party dedicated to getting rid of nuclear energy? In my own field of genetics, where there are extraordinarily exciting advances that promise to underpin the whole future approach to prevention and treatment of disease, there are serious concerns raised by, for example, the issue of what to do with information on inherited susceptibility to disease. The scientists involved must explain the underlying issues and what can be done, so that the ethical and social questions can be widely discussed by an informed public.

There is perhaps no area of public understanding that creates more difficulties than that of understanding risks and statistics, and yet millions bet on the pools and horse races showing an intuitive grasp of the odds of a bet at 5 to 4 on, or the possible rewards of an "accumulator" bet covering a succession of races. But when it comes to a balanced assessment of, say, the risks of vaccination, or of balancing the advantages and disadvantages of the risks associated with nuclear energy compared to the burning of fossil fuels, the problems appear to be much greater.

In explaining science to the lay-public, it is important not to underestimate people. We must aim to help politicians understand that science has its limitations as well as its promise; to persuade investment bankers that many science based projects are necessarily of a long-term nature; to help civil servants weigh up the costs and benefits of new methods of agricultural production and alternative sources of energy; to convince industrialists that it is necessary to support both basic and applied research, since the former is the ultimate source of the latter; and, at a local level it is important to gain the support of the community for scientific enterprises, instead of apathy, or worse still suspicion.

Improving the public understanding of science needs the support of the wider community of scientists talking to and writing for a wide range of audiences. Locally your newspaper or radio station will often be glad to hear from you. You could give talks at schools to help them cope better with their science teaching in the context of the new national curriculum and, at the same time, help your own organisation with its recruitment. Local women's and other groups will welcome the chance to learn something of your particular area of science. Looking further afield, there are opportunities in the major newspapers and journals, national television and radio, public lectures and books representing science to a wide audience. This book is packed with useful ideas about making the most of all these opportunities.

Science and technology will be the basis for all future improvements in our health and prosperity. But to achieve these benefits to their maximum we must have a public that is ready to accept and exploit the fruits of science and that will depend to a large extent on you the scientist helping to improve the public understanding of science.

Walter Bodmer, FRS
Author of the Royal Society report,
The Public Understanding of Science, (London, 1985)

Authors' acknowledgements

This book was developed from notes which we wrote to accompany a 'Communicating Science' weekend training course, which was held in the Department for External Studies at Rewley House, Oxford, in February 1989. The course benefited greatly from the financial support of the British Association and of the Oxford Trust, which funded the publication of the course notes. We are grateful to the British Association, the Oxford Trust, Geoffrey Thomas and the staff of Rewley House, and the tutors and participants of the 'Communicating Science' weekend for their contribution to this work. We also thank Geoffrey Thomas for allowing us to publish his Newton news stories, which appear on page 13.

Many people have helped us write this book, and we thank our friends and colleagues in science and in the media, including our three referees, for sharing their experience and expertise with us; the book is much the better for their contributions. Special thanks for are due for various reasons to Sir David Attenborough, Jenny Campbell, Pam Clarke, Chris Collins, John Durant, Richard Fifield, Colin Gregory, Michael Kenward, Nigel Levy, Bruce Lewenstein, Steve Miller, Jane MacRae, Richard Paul, Sean Pidgeon, Romesh Vaitilingam and Paul Viragh.

Jane Gregory received a COPUS award from the Committee on the Public Understanding of Science to cover the cost of her research for this book; we are grateful to the Committee for its support, and particularly to its Chairman, Sir Walter Bodmer, for his Foreword.

This book was written by two people in four different cities on three different continents. Its existence is very largely due to the patience and good sense with which our publisher at Longman coordinated and guided our efforts. We are very grateful to him.

Michael Shortland, Sydney
Jane Gregory, London
September 1990

Acknowledgement

Longman Group U.K. Limited are indebted to the author's agent on behalf of The Executors of the Estate of G. P. Wells for permission to reproduce an extract from *The Science of Life* by H. G. Wells.

I. *I*ntroduction

*I*ntroduction

This book is for scientists who want to improve their communication skills. On the whole scientists are no better and no worse at communicating with the public than any other group of highly qualified specialists, but they now face a new and urgent challenge: the public is suddenly very interested to hear what scientists have to say. In the past, some scientists' attempts at communication have turned out like a garden cultivated by neglect: the flowers are in there somewhere but the public has to fight its way through a tangle of weeds in order to see them. The public's need for accessible, succinct and reliable information means that scientists must cut a way through the tangle and keep the paths clear. Like anyone else, scientists can learn to communicate effectively, and we hope that this book will help them to do so.

Every year more and more books appear which are designed to enable people to communicate. Much is claimed: you can 'Conquer Your Fears', 'Develop Your Energies' or 'Talk Well and Influence People'. The problem with books like these is that the reader is required to develop this, enhance that, or embrace some mystic other-worldliness in order to benefit. Every book is written with a readership in mind: this one is for busy people who simply don't have time to fight their way through a fog of isms and ologies in order to learn how to improve their communication skills.

Other more straightforward books have been written about the practicalities of communication; together, they seem to offer every instruction a person may need, at least in matters that can be pinned down exactly enough, and sternly describe rules to be followed and laws to be obeyed. The trouble with rules is not only that they are forbidding and have an alarming tendency to multiply, but also that they tempt us to put too many things into neat, water-tight compartments. Often, there is no precise, single reason why an article or talk is not quite right. Another reason why we haven't engraved in stone any rules for writing and speaking is that our language is supremely flexible. Some of the rules we are taught at school turn out not to be rules at all, though many think them sacrosanct. One is the injunction never to split an infinitive; to do so, we are warned, is to appear to all and sundry as illiterate. Yet a sentence like 'the modern scientist fails completely to acknowledge . . .' is ambiguous. Does the scientist completely fail to acknowledge, or does he fail to completely acknowledge? 'The modern scientist fails to completely acknowledge . . .' saves the reader from having to guess by putting 'completely' in an unambiguous position – in the middle of the infinitive. Generations of children have been taught never to end a sentence with a preposition: 'a preposition is a word you must not end a sentence with'. Now it is recognized that the stress which falls on the last word of a sentence can be used effectively if it lands on a preposition. As Winston Churchill once said, grammatical pedantry is something up with which we need not put.

The dissatisfaction with your own efforts to communicate may be no more than a vague feeling born of listening to an inspiring speaker or of reading a popular science column in the newspaper. Or it may be more solid and mundane: a box of unpublished manuscripts, a pile of rejection slips. Sooner or later, you are bound to ask 'where am I going wrong?' Then, if ever, this book can help.

If your shortcomings are, like those of many speakers and writers, simply technical ones, our advice on what might be called the 'mechanics' of communication will help. But in many respects communicating is like driving a car: knowing how the car works is useful, but plenty of good drivers couldn't tell you the difference between a piston and a pump, and many skilled communicators would be hard pressed to distinguish the work of the colon and the semi-colon. What drivers and communicators need is experience and practice – and a quality which driving instructors call 'road-sense'. If there's an equivalent for science communicators, then that's what this handbook is designed to encourage in you. Road-sense involves knowing where you are and where you are going, and considering the needs of other road-users. The same applies to effective communication. A piece of writing can always be revised, cut and polished, but if the author has nothing to say, is unclear about what to say, or has no regard for the needs of his readers, then the chances are that no amount of fine tuning will make the writing road-worthy.

Communication is not a science: there is no single 'correct' way to get your message across. What is required is practice, and that means hard work, making mistakes, recognizing your mistakes and putting them right. Nevertheless, every successful practitioner knows that in some respects communication is like science: speaking and writing effectively about science offers the challenges, rewards and enjoyment which science itself can bring. We hope that this book will encourage you to apply yourself to the task with enthusiasm; there is, you will soon discover, much satisfaction to be had from communicating science to the public.

1.1 Why Communicate?

The *New Scientist* once published a cartoon of a group of scientists enjoying a party in the Senior Common Room. Only one person is unhappy: a dejected man stands on his own in a corner, while the other party-goers whisper about him. 'Who's he?' asks a guest. 'Him?' replies another, 'he's just a popular scientist'.

It is still the case that some scientists look down on colleagues who 'go public'. They give a number of reasons: if a scientist has something to say, he or she should write it up in the proper manner, submit it for peer review and then wait a year for it to be published; a medium as trivial as television is no place for something as important as science; scientists should be self-deprecating and dedicated to their work – they should have neither the time nor the inclination to blow their own trumpets; if anyone's going to speak it should be the boss; scientists who get on the radio aren't always the best scientists, and the public deserves the best; the rewards of a media career can compromise scientific integrity . . . Being resented by colleagues is a big problem in a profession where the value of your work and the progress of your career are judged by your peers. The need to conform is drilled into scientists at an early stage of their training. A good scientist is disinterested and objective – he or she has no personal stake in or feelings about the work in hand – and is completely committed, intellectually and in time, to research. Even scientists who have their colleagues' support can feel guilty about neglecting this commitment in order to communicate science.

Becoming unpopular with colleagues is not the only hazard of communicating science. What if the public commandeer your story, and distort it? The first big science story to feature in the new, cheap, popular papers was the discovery of X-rays, which broke early in 1896. Because of the press attention, Wilhelm Röntgen became famous overnight – and was extremely annoyed about it. To his great distress the aspect of his work that attracted most attention was the most superficial: the X-ray photographs. While physicists wondered about the nature of the rays, the public wanted only the bare bones: even the future King George V had his hands X-rayed, and the result was printed on the cover of the *Illustrated London News*.

And what if the public gets the wrong end of the stick? A by-product of the search for quarks was the discovery of the antideuteron in 1965. Physicists were expecting that they would one day come across this particle, which consists of nothing more exotic than an antiproton attached to an antineutron. The press, however, were fascinated by the idea of anti-anything, and journalists were eager for details. In an unguarded moment, researcher Leon Lederman told the man from the *New York Times* that this was 'the final proof that anti-people could possibly exist'. The story made the front page, and a number of readers wrote in to ask how they could locate their anti-selves.

So there are problems with communicating science: the pessimists' worst possible case is that your colleagues will never speak to you again, your reputation as a serious scientist will end up in tatters and the source of your research funding will one day mysteriously run dry. That's a high price to pay, especially when you can never be quite sure what the public will make of it all. In the past, the only scientists who could afford to run this risk were emeritus professors with a Nobel prize or two under their belts – scientists with no teaching commitments, no research contract to fulfil, and unassailable scientific reputations. Such scientists are few and far between – and not always the best communicators. So why do scientists go public? Why did Benjamin Franklin write for the *Pennsylvania Gazette*? Why did Michael Faraday give public lectures? Why did J.B.S. Haldane edit a newspaper? Why did Richard Feynman make television programmes? Why did Stephen Hawking write a best-seller? Why did you buy this book?

The answer is that communicating science can be a rewarding, pleasurable and valuable occupation, and not nearly as hazardous as the pessimists would have you believe. It's still true that only a minority of scientists consider communicating their work to the general public to be part of their job; and that few dream of addressing crowds from the top of an ivory tower on the subject of mitosis or X-ray diffraction. But scientists know that the scientific enterprise needs public support and approval, and even the stuffiest are now beginning to admit that communicating science – even if they wouldn't do it themselves – is something that needs to be done. More and more scientists are beginning to see the merits of explaining their work to the public, and are being invited to do so. You may have been asked to talk to an adult education class, issue a press release, review a book or write an article for a magazine. Perhaps your job does require you to communicate with the public, whether you like it or not. Maybe you've just decided that the costs are outweighed by the benefits of sharing with the public your knowledge about the world in which they live, and for enabling them to make judgements about the scientific issues which affect their lives.

The comprehensible scientist

> I expected [the publicity] would totally destroy my scientific career – not because I expected to get out of research, but because the average scientist is basically toilet-trained to the point where if what he does is comprehensible to the general public, it means he's not a good scientist. That's what I thought. I was wrong.
>
> Paul Ehrlich, biologist
> Author of *The Population Bomb*

In 1782 the American statesman James Madison wrote that 'a popular government without popular information or the means of acquiring it, is but a prologue to a Farce or a Tragedy or perhaps both. Knowledge will forever govern ignorance: and a people who mean to be their own Governors must arm themselves with the power which knowledge brings'. In an increasingly technological society, public knowledge of the facts and values of science is certainly a prerequisite for many democratic decisions. As well as decisions made at the ballot box, many of the choices people make every day on a personal level require some scientific knowledge: what food to buy, how to travel, how to heat their homes, or how to cope with illness. Without information we cannot sensibly make even these comparatively simple decisions, so that denying the public scientific information is to deny them their democratic right to make informed choices about the way they live their lives.

Unless they are well informed, even the most interested of the public will be unable to form opinions about the nature and value of science. People are usually good at making sense of information when it is given to them – or at least making the best of it, for often the information is confusing, biased, contradictory or incomplete. An informed public can be a great asset both to the scientific community and to society as a whole. Ignorance often leads to fear, and fear is counter-productive for an enterprise which relies heavily on public funds, both from taxpayers and consumers. Ignorance also leads to the prejudice that classes all scientific work alongside the least responsible. As the science writer Isaac Asimov observed, 'without an informed public, scientists will not only no longer be supported financially, they will be actively persecuted.' In the debate in the seventeenth century among the members of the early Royal Society on what constituted proper scientific practice, it was decided that good science was done in public, and the results should be available for all to share. Alchemy – practised in secret and reported in code – slipped another few rungs down the ladder of scientific respectability. Three hundred years later, the Royal Society's 1985 report urged scientists to contribute to the task of informing the public: 'It is clearly part of each scientist's professional responsibility to promote the public understanding of science'.

There are many economic and social benefits of wider scientific knowledge. We bear the enormous cost of treating avoidable diseases, cleaning up after unwitting polluters and providing crash public education programmes in times of crisis. Much of the information distributed after the Chernobyl accident was basic facts about radioactivity, and much of the information about AIDS was basic facts about the nature of viruses. Manufacturers need to train workers and to inform consumers. Any nation that relies for its prosperity and security on science and technology-based industry and services needs a plentiful supply of scientists and a public which supports the scientific enterprise.

Science is part of our culture and heritage, and scientific knowledge ought therefore to be common property. No poet would insist that his work be read only by other poets; nor would any actor forbid all except actors to watch his performance. There is great pleasure to be had from understanding nature – this is surely one of the main reasons why people do science – and there is no reason why scientists should not share this pleasure with the public. The astronomer Carl Sagan wrote: Understanding the world is a kind of joy, and I find that every time people, ordinary people, understand some aspect of nature they hadn't grasped before – why the sky is blue, why the Moon is round, why we have toes – they are delighted. This is a delight first in the joy of knowledge itself and second because it gives them some intellectual encouragement: they discover they're not so dumb as they had been told they were.

Brain food

> The true and specific function of popularisation is purely and simply to introduce the greatest number of people into the sovereign dignity of knowledge, to ensure that the great mass of people should receive something of that which is the glory of the human mind . . . to struggle against mental starvation and the resulting underdevelopment by providing every individual with a minimum ration of spiritual calories.
>
> Jean Rostand, biologist

Some popular science is written by people who claim to be experts, but whose work wouldn't stand up to peer review. The only place it could be published is the popular press, and published it is. Legitimate scientists, uncomfortable about appearing alongside such people, are content to let their work reach a select few through the pages of the specialist press – thereby leaving the public press open to their less respectable counterparts. If you have published an academic paper and its conclusions are of interest or consequence for the public, then your job is only half done – your work should be in the newspaper or on the radio as well. As the chemist F.W. Clarke wrote in 1872, 'there is a demand for science, or the trash which is written would not be read.'

In the late nineteenth century the popularizers were scientists themselves. During the twentieth century the task passed to journalists, and many scientists gave up on the press because of what they saw as irresponsible reporting. This produces a vicious circle: a scientist is worried about inaccurate reporting, and doesn't co-operate with the press. The press does the story without the scientist, and so the story is more inaccurate than it might have been. The scientist gets even more worried about the inaccuracy of the press, and further resolves never to co-operate.

In a recent study of the popularization of science, John C. Burnham looks at what he sees as the battle between science and superstition for a place in the public mind. He concludes that, for the time being at least, science has lost – because scientists stopped trying to communicate science to the public.

Science was commandeered by anyone who chose to make scientific statements, irrespective of whether they had the authority or expertise.[1] Burnham has a point: after all, although there are political correspondents and sports writers we still hear from the politicians and athletes themselves, and we know who they are and why they have the authority to speak. Science stories can't include the words of scientists if no scientist is prepared to speak. Such stories read like an account of a football match which tells you the score but not which teams were playing.

On a personal level, communicating science to non-scientists is an excellent way of getting your ideas straight in your own mind. In 1860 the evolution debate was hotting up, and Darwin's friend Thomas Huxley knew that he would soon be called upon to defend Darwin's ideas. Huxley was due to give a series of public lectures and chose as his subject the relationship between man and the animals, because 'some experience of popular lecturing had convinced me that the necessity of making things plain to uninstructed people, was one of the very best means of clearing up the obscure corners of one's own mind.' The chemist Irving Langmuir has taken this argument one step further, and said that if you can't explain your work, you can't claim to understand it. 'Anyone who can't explain their work to a fourteen year old', he said, 'is a charlatan'.

If your motive is money, think again. Some writers – mostly of textbooks and fiction – do make money, and a larger number write in the hope of getting rich, but most authors know that there's no money in writing. There is some glory to be had, and the shift in attitude of the scientific establishment over the last few years has produced some encouraging results. Anyone who doubts whether communicating science to the public is a suitable occupation for a scientist should note that the eminently respectable Royal Society has recently added to its list of prestigious prizes the Michael Faraday Award, which is given each year to the scientist who has made the most significant contribution to the public understanding of science. Other learned societies and professional institutions are making similar moves to encourage science communication: more information is given on page 171.

What's left, apart from the requirements of a job and a sense of duty? Part of the answer is simply egotism, and personal ambition is nothing to be ashamed of. Mind you, there's nothing like a visit to a public library to teach you something about the vanity of human hopes. Samuel Johnson wrote: 'who can see the walls crowded on every side by mighty volumes, the works of laborious meditation and accurate enquiry, now scarcely known but by the catalogue, without considering how many hours have been wasted in vain endeavours!' Nowadays, things are not a great deal better if you intend to break into the book-publishing world. But there are plenty of other ways to make your mark – articles, talks, interviews and news reports can all satisfy another major motive for communicating: propaganda. The word has unpleasant connotations, but its literal meaning is simply the dissemination of information. You have interesting ideas or discoveries and want to bring these to the attention of the world at large – you want to push opinion in a particular direction. You want to persuade people.

The persuaders

> The missing element in much of scientific literature is the unwillingness of authors to accept as an essential part of their self-imposed task that they should actively seek to persuade potential readers of the interest of what they have to say. A description of the reasons why a piece of research was undertaken and an assessment of its importance (which may frankly be subjective) are obvious ingredients of a persuasive article. So too may be a little enthusiasm.
>
> John Maddox, Editor, *Nature*

Some people think the job of scientists is simply to present information objectively, but history shows us that the majority of scientists, past and present, have tried to use information to change opinion. Charles Darwin's *Origin of Species* is a classic of science; modern biology stems from it. If you read it today, you are bound to be struck not only by the wealth of evidence Darwin has brought together, but also by the manner in which he presents it. His tone is restrained and his style is discursive and anecdotal – but this, like his careful use of figures of speech, was calculated to win over his Victorian readers. Darwin's book was, he claimed, 'one long argument'. His theory of evolution did not triumph solely because he used language effectively, but there is no doubt that the persuasive style of the *Origin* contributed to the speed at which Darwin's ideas became both widely known and widely accepted.

Some people think that good communicators have to struggle to efface their own personalities, and conform to an out-dated stereotype – the serious, pulseless automaton subjecting Nature to the unassailable scientific method. No scientist would claim that this character exists, yet many seem to imitate him when they speak or write for laypeople – they lose their personality and enthusiasm for science, and spout facts with all the subtlety of an exploding encyclopaedia. Some people hide behind the stereotype, and use it as an excuse for keeping themselves to themselves. Other people think that showing their feelings is 'acting' or 'pretending', and that pretence has no place in science, which thrives on the honesty and integrity of scientists. But communicating is not about pretending: we are not actors with script-writers and directors. Communicating is about being honest, about owning up to feeling good or bad or disappointed or excited or even just interested in science. Too many people who talk about science give the impression that they couldn't care less about it, and if that's the case then there's no reason why anyone else should either.

There is immense enjoyment to be had from communicating science. You have a fascinating piece of experience to convey and there is great personal pleasure to be had from conveying it well. Writing well, or communicating in any medium effectively, is as satisfying as any other important, stressful, difficult and tiring job well done. That, finally, is what is going to keep you going for hour after hour of rewriting, rehearsing and filing page after page in the rubbish basket, until you finally produce something of which you can be proud.

Any journalist will tell you that what people find interesting is people. Even the most astounding scientific ideas have benefitted from a little personal support: many people first became aware of relativity not because it revolutionized physics, but because it was brought to us by one of the great characters of the twentieth century. Einstein was happy to share his enthusiasm with the public, and his popular book on relativity is still the best. We can't all hope to make a contribution to science as profound as Einstein's, but you don't need to be a scientific genius to be a good communicator – all you need is to have the science straight in your own mind and know how to get your message across. All scientists, whatever their job or scientific achievements, can contribute to the enormous task of communicating science to the public.

Recommended reading

David Evered and Maeve O'Connor, 1987, *Communicating Science to the Public* (Chichester: Wiley/Ciba Foundation). Essays and discussion from scientists and communicators in a variety of media.

The Royal Society, 1985, *The Public Understanding of Science* (London: Royal Society). The report which prompted much of the recent British activity in this field; it includes a number of recommendations and a useful bibliography.

Reference

1 John C. Burnham, 1987, *How Superstition Won and Science Lost: Popularizing Science and Health in the United States* (New Brunswick: Rutgers University Press).

1.2 Science and the Public

There was a time, half a century or so ago, when a speaker or writer could take a great deal for granted. British scientists talking about their work could assume that nearly all their listeners were male, that they came from families that were more prosperous than the national average, and that they shared a common culture: Church, Empire, and a Classical Education. Happily, things have changed remarkably since then. The general public is more literate, and more scientifically literate, than before, and your audience will consist of people with a wide range of backgrounds and interests.

In the early part of the century authors and speakers tended to sound alike whether they were dealing with art, architecture or anatomy. Even though each discipline had its jargon, a scientist could still write in much the same style as anyone else. In 1920, *Nature* published this:

> Cream plastid colour is recessive to white, and deep sap colours are recessive to pale. When a cream-eyed strain lacking the pale factor is bred with a white-eyed plant of some pale shade, the four possible combinations appear in F_2, but not, as we should expect in the case of two independently inherited characters, in the proportion 9:3:3:1, with the double recessives as the least abundant of the four forms.

Were this author working today, she might have written something like this paragraph from a recent issue of the same magazine:

> To test whether T cells expressing the 2C TCR would arise in comparably high numbers in a non-autoreactive H–2 environment apart from H–2b, transgenic H-2S animals were generated by backcrossing founder P twice to (C57L x SJL) F_1 mice.

Something important has happened during the seventy years or so that separate these two paragraphs. Scientific language has diverged from the mainstream of literary language and divided into a large number of small, winding tributaries. The second passage also presents familiar words – 'expressing', 'environment', 'generated' – in an unfamiliar context. The last scientific representatives of the old order who wrote on almost every topic in the same style were authors such as Bertrand Russell and J.B.S. Haldane. Their work can still be enjoyed today, but it bears a distinctly old-fashioned air reminiscent of top hats and starched collars.

Different styles are used for different purposes, and part of the communicator's skill is to find the most appropriate: a style which suits both the subject and the audience. Imagine how the results of Newton's *Principia Mathematica*, published three centuries ago, would be reported in today's newspapers. Geoffrey Thomas has suggested that the breakthrough might well appear in the *Daily Mail*, *Daily Express*, and *Sun* like this[1]:

COME OFF IT ISAAC

In a book just published, Isaac Newton, a Cambridge mathematician, claims that a body will remain at rest or continue in uniform motion in a straight line unless acted upon by an external force. Remain at rest, undoubtedly. But move without being pushed or pulled? Never!

This is the worst sort of nonsense we've heard in many a long day. Come off it Isaac, the great British public can't be fooled as easily as that!

WHAT GOES UP MUST COME DOWN

In a book out recently, Cambridge professor, Isaac Newton, 45, unmarried, gives an explanation for the well-known tendency things have to fall when you let them go. This account, which is too technical for us to cover here, also apparently explains why the world high jump record stands at only 2 feet 11 inches.

Who said British science was dead?

Isaac Newton, who is said to have turned down a lucrative offer from Berkeley, said he was over the moon about his theory. We say: put this man in charge of the Royal Mint.

WHAT A SHOCKER!

Are you attracted to your neighbour's wife, or your neighbour even? Feeling guilty about it? Don't.

British boffin Ike Newton says it's perfectly normal. He says it's a law of nature for everybody to be attracted to everybody else.

And what's more, the closer you are together the stronger the attraction you'll feel.

We knew it all along, although we still find some bodies more attractive than others – see saucy Karen on page 3.

Celebrate with the *Sun* – don't miss our great Principia T-shirt offer.

Catching the idiom and style of your likely reader or listener is important. This doesn't mean that your own style will inevitably become a pale imitation of someone else's – you will simply become attuned to what is appropriate for a particular theme and a particular readership.

The relationship between science and the public is reflected in the way scientists address their audience. Popularizers have had different intentions: in the middle of the nineteenth century they wanted to bring to the masses the joy and moral benefit of knowledge; they wanted to reveal the hand of God in Nature; they wanted, by exposing the world as an organized, ordered system, to keep the working class in its place. The workers had other plans: they turned enthusiastically to night-schools and popular books because they recognized the power that knowledge brings. Magazines were cheap and widely read, and there was much science in the general press. Parents wanted their children to learn science, and there was a huge market for children's books. The science in them was self-improving, both intellectually and economically. In one children's book young George Brown's father is not angry when George uses a magnifying glass to burn a hole in a friend's coat, because 'Mr Brown worked in a factory; and he was rather glad to see that his son was thinking about other things than throwing stones and kicking out his boots .'[2]

The newspapers took science seriously: the lectures of Thomas Huxley, Louis Agassiz and Asa Gray were often published, and in 1872 a special edition of the *New York Tribune* which contained the text of physics lectures by John Tyndall sold 50 000 copies. Science stories had previously been written mostly by scientists, but as science grew more popular, the task of reporting it fell more and more into the hands of reporters. This resulted in some inventive writing: the public interest in radium led to a number of speculative stories, including one which suggested that feeding radium to chickens could result in eggs that would boil themselves. However, scientists kept a close eye on the popular press, and when the need arose they rushed in to set the record straight.

By the end of the nineteenth century, science journalism reflected the growing division between those who felt that science was the answer to all our problems, and those who felt that it might be causing some of them. For the press, the debate was resolved by World War I. Fought with gas and aeroplanes and won by tanks, it demonstrated science's contribution to the power and status of nations. Science was back on its pedestal.

The economic boom of the 1920s had positive consequences for popular science. Press magnate Edwin W. Scripps launched the Science Service: a news agency which offered 'drama and romance . . . interwoven with wondrous facts', and claimed that 'drama lurks in every test tube'. By the mid-twenties, one scientist remarked, 'science has at last become articulate, not to say garrulous'. Science may have been garrulous, but it had plenty to say: the 1920s brought passenger aeroplanes, arctic explorations, mass production, Freud, Einstein and radio to public notice, and closed with the discovery of Pluto.

The Depression of the 1930s brought a dip in the coverage of science, and during the War the papers dealt with more pressing matters. Coverage increased again through the 1940s: the morale-boosting, optimistic rhetoric that carried readers through World War II persisted, and the accelerated technological achievements of the war years were celebrated in similarly upbeat terms. The launch of Sputnik in 1957 astonished the world; it prompted the reorganization of science education, particularly in America, and the 1960s saw a huge increase in the number of people who learned science at school. This led to a more informed public, but one which had the ability to criticize as well as support. The impetus for many of the protest movements of the late 1960s and early 1970s, particularly those concerned with the environment and nuclear weapons, came from young people exercising not only their political awareness but also their new-found scientific knowledge. The new environmental science appeared often in the media, but much of the coverage was critical of technology and the other sciences. In 1930 William L. Laurence, science reporter for the *New York Times*, had written 'True descendants of Prometheus, the science writers should take fire from the scientific Olympus, the laboratories and universities, and bring it down to the people.' By the 1960s science journalists no longer saw themselves as missionaries, but were beginning to feel that the criticism and commentary that other reporters offered their readers should also be part of their science coverage. John Lear of the *Saturday Review* wrote that 'The spirit of untrammelled inquiry and skepticism required of journalism in other fields must become a standard in science writing.' The journalists' job is now not simply to protect and promote science, but to report it just as they do politics or sport – they report, but they also analyse, criticize and draw conclusions.

But what of science and the public now? Recent survey research has produced information about current public attitudes to and understanding of science and technology.[3] In order to measure what the public actually knows about science, researchers devised a series of questions about various scientific phenomena. Most of the scores were similar on both sides of the Atlantic. In some areas, the public knows its science. Almost everybody knows that hot air rises, and 86% know that the centre of the Earth is very hot. 94% know that sunlight can cause cancer, 72% that the continents are moving about very slowly on the surface of the Earth, and 60% that the oxygen we breathe comes from plants. Other areas are more of a mystery. Only 31% know that atoms are smaller than electrons, and so the other 69% must have very little understanding of the nature of electricity, despite its huge presence in their lives. 55% think that antibiotics kill viruses as well as bacteria, which must affect their judgement of the success or otherwise of medical practice. The nature of the distinction the public make between 'natural' and 'artificial' is revealed by this statistic: 70% think that natural vitamins are better for them than vitamins made in a laboratory.

Words of wisdom

Scientists are so blamed wise and so packed full of knowledge . . . that they cannot comprehend why God has made nearly all the rest of mankind so infernally stupid.

Edwin W. Scripps
Newspaper magnate

Surprisingly perhaps, 30% of the people asked thought that the Sun goes round the Earth. Of the people who knew that the Earth goes round the Sun, only 34% knew that it takes a year to do so. There are always influences from outside science which affect what people believe: while 80% of British people believe that humans developed from earlier species of animals, only 45% of Americans agree with them.

When asked what it means to study something scientifically, only 3% mentioned the construction of theories and only 10% mentioned experimenting. However, when presented with a specific problem – in this survey people were asked to choose from a list a method for working out why a drug wasn't working well – 56% chose a controlled experiment.

Other work has shown that scientific information does not arrive to take up a pre-ordained place in empty minds – it finds, if it is appropriate, a niche among the knowledge, misconceptions and beliefs which already fill the minds of even the most ignorant. Nor do the public accept uncritically anything the experts throw at them: they think carefully about what they know of the expert's background and vested interests. Scientific knowledge is not received as disembodied absolute facts: it is seen as a product of a world of real people in real situations, and is integrated into everyday life.

New Scientist regularly surveys public attitudes to science and scientists. In 1975 the magazine asked respondents to describe scientists, and many people named actual scientists in support of their descriptions. The most frequently mentioned scientist was one who had been dead for over 2000 years: people thought that scientists were like Archimedes. In second place came a fictional character, Professor Branestawm – the archetypal mad scientist who lives in a tangle of complicated mechanical devices which make him tea and answer the door. Third in the poll came Jacob Bronowski, whose television series was a great success. Fourth was Marie Curie, and she was followed by Darwin, Einstein, Faraday, Fermi, Fleming and Galileo. Another *New Scientist* survey conducted ten years later shows just how quickly things change: Archimedes had dropped out of the charts, and Einstein was in first place. Newton, who had been in thirteenth place in 1975, came second in 1985 – he was on the pound note at the time. Marie

Curie, who has become associated with the positive consequences of her work, joined other medical pioneers Fleming and Pasteur in third, fourth and fifth place. The rest of the top ten was an electromagnetic consortium of Bell, Faraday, Marconi and Baird, who were joined by Barnes Wallis, scientist and hero of World War II. The only living scientist in the 1985 poll was the electronics inventor Sir Clive Sinclair, who came sixteenth – one place behind the electrical inventor Thomas Edison.

Current public attitudes to science are encouragingly positive. Averaging similar British and American survey results shows that 90% think that governments should fund blue-sky research, and 86% think that science and technology are making our lives healthier, easier and more comfortable. Americans are apparently happier with 'progress' than Britons: only 37% of Americans but 60% of Britons think that science makes our way of life change too quickly. This finding was confirmed by two others: more Americans than Britons thought that animal experiments should be allowed, and that more jobs would be created than lost as a result of computers and automation. However, overall there was a slight majority against animal experiments, and a slight majority disagreed that computers and automation would create jobs.[4]

Assessing levels of public interest in current events, surveys repeatedly find that the public is very interested in science. One survey found that while 77% of people asked were interested in local issues, 57% were interested in defence and 59% were interested in business news, 92% were interested in medicine and 88% were interested in science.[5] In 1988 two thousand British adults were shown a list of thirteen newspaper headlines, and asked which they definitely would or wouldn't read. Looking at the headlines, can you tell which attracted the most attention?

Heathrow robbery and drugs deal
EastEnders star in baby shock
New clue in hunt for AIDS cure
Robots for housework soon
FA cup winners surprise
Can pesticides in rivers be controlled?
Government backs massive investment in science
UN calls for British action against South Africa
Astronomers discover new galaxy
60's pop idol in comeback
'Was Darwin right?' say scientists
Heart disease our own fault claim experts
New government cuts announced

According to the public, the most interesting headline in this list is the one about heart disease. Second came the government cuts, and third work on

AIDS. Pesticides were slightly more popular than astronomers – they came fifth and sixth respectively – and investment in science came eighth, while Darwin scrapes home in ninth place.[3] All in all, science holds its own surprisingly well against competition from soccer, sex and soap.

Clearly people are aware that science is important and relevant to their lives. However, when asked whether they were 'informed', the percentage of people who claimed to be informed about science was much lower than the percentage who claimed to be interested in science, while for sport and politics the figures for informedness and interest were about the same.[3] It is difficult to judge a result like this, because people's assessments of their own knowledge are influenced by many factors which have nothing to do with what they actually know. People who are aware of the responsibilities of living in a democracy might be embarrassed to confess to being ill-informed about politics, for example, and yet some people who would be ashamed to admit that they had never read Shakespeare would have no qualms about claiming to be ignorant of science. While scientists might feel sorry that the public is not embarrassed by its ignorance, there is a positive aspect to this attitude: the public is keen to know more about science, and a confession of ignorance is a useful first step towards learning more. Most people believe that increased knowledge would be a good thing: when asked to respond to the statement 'It is not important for me to know about science in my daily life', 63% of Britons and 82% of Americans disagreed.[4]

Many of scientists' feelings about the public's perception of them as people are unjustified. Scientists are generally held in rather high regard. One survey offered a list of personal characteristics, and asked the respondents to decide which of the characteristics could be attributed to scientists. It seems that the public thinks that scientists are responsible, concerned and not particularly different from anyone else. They are no more or less sociable, arrogant or eccentric than other people, but they are more secretive than average.[5] The feeling that scientists are secretive is an interesting reflection on the access laypeople have to scientists and scientific knowledge. While some scientists may need to be secretive, many will be unaware of just how inaccessible they are to the public.

So the public's attitude to science is positive, but their knowledge of science is limited. There is nothing particularly extraordinary about the levels of ignorance of science: levels of ignorance are high in politics, history, literature and foreign languages. As one researcher has pointed out, however surprised we might be about statements which provide a measure of knowledge, we should assess these in contexts such as that provided by his home state, Kentucky, where 20% of the population cannot read and write.[6] We can all be appalled by statistics like these, but the fact that the public knows no less about science than it does about a number of other important fields is no reason to be content with this ignorance, particularly since the public is keen to know more.

Recommended reading

John R. Durant, Geoffrey A. Evans and Geoffrey P. Thomas, 1989, 'The public understanding of science.' *Nature*, **340**, 6 July, 11–14. This first national survey of British public understanding of science produced some startling data, and presents it in comparison with the corresponding data from the US.

William Hively, 1988, 'How much science does the public understand?' *American Scientist*, **76**, September/October, 439–44.

Michael Kenward, 1989, Science stays up in the polls, *New Scientist*, 16 September, 39–43. A report of a national survey in Britain which focuses a public attitudes to science, science policy and science funding; it includes a brief survey of Australian attitudes towards science

References

1 Geoffrey Thomas, 1987, Paper presented at a conference to mark the tercentenary of Newton's *Principia,* Department for External Studies, University of Oxford.

2 W. J. Pope, 1893, *Pope's School Readers: Standard II* (London: Smith, Elder & Co), p. 7.

3 John R. Durant, Geoffrey A. Evans and Geoffrey P. Thomas, 1989, 'The public understanding of science.' *Nature*, 340, 6 July, 11–14.

4 Geoffrey Evans and John Durant, 1989, 'Understanding science in Britain and the USA'. *British Social Attitudes: Special International Report*, edited by Roger Jowell, Sharon Witherspoon and Lindsay Brook (Aldershot: Gower), pp. 105-20.

5 Michael Shortland, 1987, 'Networks of attitude and belief: science and the adult student.' *Scientific Literacy Papers*, edited by Michael Shortland (Oxford: Department for External Studies, University of Oxford), pp. 37–51.

6 Robert K. Fullwinder, 1987, 'Technological literacy and citizenship.' *Scientific Literacy Papers*, edited by Michael Shortland (Oxford: Department for External Studies, University of Oxford), pp.31–5.

1.3 How not to become an Expert

There are two types of expert. The first conforms to the dictionary definition: experts are people having special knowledge or skill. The second, who are Experts with a capital E, are found not in the dictionary but in the media. These Experts are people of unknown qualifications and hazy background who pronounce on every subject with equal confidence and in the patronizing tones which make it quite clear that they know a whole lot more than the ignorant masses for whose benefit they speak. As a scientist you are already an expert – the problem for science communicators is how not to become an Expert.

Avoiding Expert-ese

We've all come across scientific Experts. The gulf between scientists and the public is exaggerated by the way in which they talk or write – like all-powerful authorities casting pearls of wisdom before the ungrateful masses. When you are invited to speak or write about your work it's easy to slip into Expert-ese, and to use language in a way which suggests an attitude towards the public which may not exist. For instance, terms like 'obviously' and 'of course' give the impression that you expect people to know what you are going to say, and those who don't may feel uncomfortable. The same goes for phrases like 'this section is very simple' and 'it is interesting to note' – readers may not agree.

Some scientists think that when they write for the public they are 'bringing it down to their level', or 'taking a few steps backwards' in order to make things clear. If you think you are taking science 'down' or 'back', you might as well give up now – however hard you try, you are bound to end up sounding patronizing or condescending. There is a kind of natural selection at work in popularization: your work will survive if it is adapted so that it is appropriate to your readers.

Deciding what your audience already know and think is difficult, but it helps to separate knowledge from intelligence. The public may be completely ignorant of your subject, but are probably quite capable of understanding it if you explain yourself carefully.

Intelligence and knowledge

The reader for whom you write
is just as intelligent as you are but
does not possess *your* store of knowledge,
he is not to be offended by a recital
in Technical language of things known to him
(e.g. telling him the position of the heart and lungs
and backbone).
He is not a student preparing for
an examination & *he does not want to be*
encumbered with technical terms,

> his sense of literary form & his sense of humour is probably greater than yours.
> Shakespeare, Milton, Plato, Dickens, Meredith, T.H. Huxley, Darwin wrote for him. None of them are known to have talked of putting in 'popular stuff' & 'treating them to pretty bits' or alluded to matters as being 'too complicated to discuss here'. If they were, they didn't discuss them there and *that was the end of it.*
>
> H.G. Wells to Julian Huxley while they were writing *The Science of Life*

Some people – even some scientists – think that science is much more difficult than any other subject. To be a scientist, or to understand science, you have to be really clever. Whether this is true or not – and it probably isn't – when you try to communicate you must work on the assumption that anybody, whatever their intelligence, will be able to understand science if you explain it carefully enough. In 1938 the biologist Lancelot Hogben wrote that in the nineteenth century scientists had the courage and the conviction needed to communicate science because they believed that the public would understand. 'In the Victorian age big men of science like Faraday, T. H. Huxley and Tyndall did not think it beneath their dignity to write about simple truths with the conviction that they could instruct their audience . . . The key to the eloquent literature which the pen of Faraday and Huxley produced is their firm faith in the educability of mankind'. Their success suggests that Faraday and Huxley were right to believe in the capacity of laypeople to understand. If people don't understand you, it will be because you didn't make yourself understood.

The old idea that the scientist is a morally superior person engaged in a pure, objective, value-free pursuit can no longer hold its own against the historians and sociologists who, while they may concede that science is getting closer to some absolute truth about nature, will tell you that what we believe, and why we believe it, has more to do with our values, preconceptions, expectations and non-scientific ideas about nature than many scientists would like to think. When you talk to people outside the scientific community, you may find that their values are different from yours, and this may make communication difficult. This difference is sometimes difficult to judge, but it is one which needs to be taken into account, particularly if you are dealing with a controversial subject, or one of immediate public concern. It is worth thinking about:

Consequences – are there any you have chosen not to consider because you don't think they are important? Might they be important to other people? Of those you did consider, would you have got a different answer if you had approached the problem differently?

Research – who decided what research should be done, and for what purpose? When funds were allocated, which were the fields that were left out? Had these been encouraged, might the problem now look different?

Sources – how do you decide between good and bad evidence?

Experts – are your sources independent and reliable? Would the public trust them? If you have a choice of good sources, do you get differing advice from them?

Making information available about the risks as well as the benefits is important for maintaining the credibility and integrity of your institution, and for establishing public trust. It can take a lot of expensive time and effort to make up for the ground lost when laypeople find out about science's shortcomings for themselves, while your telling them can have benefits in terms of public contact, increased trust and a high profile for accurate information. A more complete picture of science must include a discussion of uncertainty, both of specific data and, on a larger scale, of scientific knowledge in general. There are undoubted problems with this: one is that public confidence is easily lost to an alternative which is presented in more certain terms, though it may not warrant them. But the advantages are far-reaching: the public has a more accurate picture of the decisions you are making on their behalf, and you are providing them with an education in the ways of science which will enable them to better understand scientific decisions.

Explaining

> Explain, explain, explain – but without resentment.
>
> Fred Jerome, Scientists' Institute for Public Information

The people who read about or listen to your work will want to know something about you. One of the barriers that has built up between the scientific community and the general public has been created by the belief that when a scientist enters a laboratory, he must hang up his emotions, temperament and feelings just as he hangs up his coat. You know that is nonsense, but the image of the scientist as a white-coated, eccentric automaton has done a great deal of harm. The more effort you make to dislodge that image by portraying the excitement, challenges, and human qualities of science, the more likely it is that ordinary people will be interested in what you have to say.

How you feel about science is just as important as what you know. For many people, their emotional response to a situation is more powerful than their considered response to the facts. In 1946 the chemist Harold Urey wrote about the potential of atomic weapons in *Collier's* magazine: 'I write this to frighten you. I'm a frightened man myself. All the scientists I know are frightened.' When in 1968 the physicist Ernest Sternglass wanted to alert the world to the dangers of fallout from nuclear weapons tests, he chose to do so through the medium of the popular magazine *Esquire*, and his article was called 'The Death of all Children'. Many media-watchers have noted how the opponents of all things nuclear have invariably won the propaganda war against the nuclear advocates. One researcher who watched congressional hearings about reactor safety in 1973 found that the advocates spent about 80% of their time giving technical details and quantitative assessments of possible consequences, and only 20% of their time discussing people's feelings or values. The opponents used the reverse strategy, and spent most of their time talking about the effect of nuclear power on people's health, feelings and peace of mind.

We can illustrate the shortcomings of the straightforward, emotion-free approach to science communication by comparing a 'late news' article about a road accident with an eye-witness account. The newspaper tells us that a child of a particular age living at a particular address was killed outside her house when she ran into the path of a car. That certainly carries information, but anyone who has ever seen or been involved in an accident knows that this is far from the whole story. An eye-witness will tell you of the little body lying in the street, the car run off the road, the shock of the driver and the agony on the face of the child's mother as she rushes into the street. The newspaper report you read in seconds and forget in minutes, but the eye-witness account will stay with you for some time, and may even change your driving habits.

In writing about science it is unlikely that you will have a story as raw and emotive as that one, but when you write about your work you are the eye-witness, not a disinterested reporter. There is no reason to suppress the emotional aspects of your work – your readers want to share in the excitement of science, as well as the facts.

Filling the knowledge gap

When communicating science to the public, your most valuable source of information is yourself. However, since very few people manage to work alone and write about themselves and still be entertaining, you'll need some others. Even if you are writing or speaking about your own scientific specialty, you will find when you write for the public that there are gaps in your own knowledge: you may know all there is to know about Linnaean taxonomy, but your readers may be interested to know more about Linnaeus. It's always worth digging up some new information: if you rely solely on your own knowledge you may find it difficult to sound spontaneous and enthusiastic about material which is so familiar to you. Thorough and accurate research adds variety, provides a sense of the community of scientific work and gives you confidence and authority; it helps you ask the right questions when you interview your sources, ensures that you don't miss out on a vital piece of information, and enables you to answer any questions that your work might prompt.

The place to start your research is with the public. What interests them? What do they know? Like much early popularization which aimed to set the public straight on scientific matters, you may have to deal with their misconceptions before you can get into the science. You won't be able to conduct a national knowledge quiz yourself, but by asking around, ringing a few friends or writing to relatives, you can find out what they think about specific points. Then you have to decide whether any misconceptions actually make any difference to what you are trying to say, and to find the information the public will need before they can understand or be interested in your subject.

Apart from the science, you have to think about what the public knows about the background to your story. If you say 'because of the problems we had here last year we've redesigned the penguin enclosure' you will find that at least half of your audience will be wondering what the problems were, since they will have forgotten that the press had a field day last July when seventeen penguins were killed by a mystery virus. You may have forgotten some of the details yourself, so go back to the papers and find out what the public were told, as well as what the scientific community decided about the episode. Cover the background succinctly, or you will find that you are retelling yesterday's news.

Journalists can get away with making mistakes – it is an occupational hazard. A scientist cannot, so be careful with proper names, dates, figures, and any facts that could be looked up in an encyclopaedia. Make notes as you go along of where your information comes from, and then go back and check it when you have finished. Then check it against another source, and then ask a colleague to cast a critical eye over it.

If a non-scientist friend asked you where he could find out about your subject, you probably wouldn't suggest he spend an afternoon in your institution's library. It won't be the most appropriate starting point for you either, if you want to write or speak for the public. There are many other sources outside your own institution which can provide different types of information to help you put together an interesting and informative story, and you should use them. Your article or talk should move from the general to the specific, so start with general sources.

Public libraries

Public libraries provide many services that academic and institution libraries can't. They are an excellent source of information about local events, people and history. A story which merited two lines in an encyclopaedia may have covered the front page of a local paper, and the library will be able to provide it. Careers which 'started small' are often followed in great detail by the media of their home town; stories headlined 'local girl wins science prize' often contain just the sort of homely, mundane and fascinating information which is missing from the appointments notice of the institute in-house newsletter. The general public is just as interested in a scientist's hobbies and home background as they are in her academic record, and by showing scientists as ordinary people you make them more accessible and friendly.

Libraries also have contacts with other local institutions which may be able to help you. Special interest groups, such as parents' associations or hobby clubs, can be contacted through the library. Libraries often have close ties with local government, and will be able to guide you to records of public inquiries and local policy.

You can save a lot of time by asking librarians. Librarians are an underrated and undervalued breed, and have far wider skills than might be apparent. They know what information is available and where to find it. They may know of indexes and catalogues which can guide you, or of archives you may not have seen. They know about local events and personalities. If your librarians can't help you, they'll know someone who can.

Photo libraries

Photo libraries are not hosts to a stream of visitors like libraries of books, but many of them welcome callers. Apart from providing pictures which you can use in your article or talk, photo libraries are an excellent source of inspiration: the science in your story may have happened in a town you've never visited, or been produced by a scientist you've never seen, and a photograph could provide you with a mental picture that you can share with the public.

There are many commercial photo libraries, some of which specialize – maybe in travel, the nineteenth century or flowers. Museums and galleries often have photo libraries which contain pictures of their collections and other material, as do some hospitals. Many large companies keep photos for advertising and public relations purposes, and they may provide copies free of charge. Broadcasting organizations and newspapers sometimes have photo libraries, and may also be able to sell you a still from a programme or a copy of a photograph by a staff photographer. If you have trouble finding a photo library, find a book with good pictures and read the acknowledgements – the source of library photographs is always printed in the book.

Most photo libraries charge a fee, and the size of the fee depends on what you do with the picture. A slide for a talk to schoolchildren may be cheaper than the same picture for your company sales brochure. Some libraries also charge for the time it takes their staff to look for a suitable picture, whether they find one or not. Most libraries will take your order over the telephone, and most – if you provide some credentials such as your affiliation or business address – will send you a selection of pictures to look at before you have to part with any money. Once you have made your choice, you will be required to state in writing that you will acknowledge the library, and observe any conditions regarding further use of the picture.

Cuttings libraries

There are some libraries which keep only newspaper cuttings, filed by subject. Often these have a postal service and will send you copies of relevant material for a fee. If there isn't an independent cuttings library in your town, the local newspaper or radio or TV station will probably have one.

Museums

It is easy to forget that a great deal of research goes on behind the scenes in museums, and that their apparently invisible staff have a great deal of expertise. Your first step should be to look at the relevant exhibits. Then go to the information desk and find out the name of the curator responsible for those particular exhibits, so that you can then contact them by name asking for specific information beyond that which is on display. If you can't visit the museum yourself, a phone call to the information desk will set you on the right track.

As well as providing information, museums often have a book shop or souvenir shop which could be a good source of pictures, or of toys and models which you could show your audience.

Professional organizations

You are probably a member of a professional organization: a trade union, specialist institute or learned society. The magazine you leave unread every month contains the sort of information that would provide someone from another field with an insight into the attitudes and concerns of your profession. There are thousands of these magazines – one for every trade – and they are very revealing. Because they are intended to be read by a closed community with shared interests they are open, forthright and directly reflect the concerns of their members. The nuclear scientist who is commissioned to write a magazine article entitled 'Chernobyl – ten years after' would benefit from reading the May 1986 issues of *Farmer's Weekly*, *The Greengrocer* and *Health and Safety at Work*, as well as the *Bulletin of the Atomic Scientists*. Trade publications are produced by an editorial office which is staffed by experienced and knowledgeable people who may be able to put you in touch with articulate and authoritative members, and they may have a library that you could use.

Special interest and pressure groups are also useful and often enthusiastic sources of information. If you can't find a particular group, ask its opponents.

Research organizations

Market research is big business. As well as carrying out surveys to determine how people will vote and whether or not they will be interested in buying cherry cola, research organizations also monitor attitudes and interests. The results of these surveys can be a great help to a writer or speaker, as they provide an instant analysis of public concerns and have the authority which many people associate with statistics. Often surveys are repeated every few years, and the historical development of an attitude or behaviour can be interesting – people are interested in change and the reasons for it, especially when they think it can help them predict the future. Market research companies and census and records offices publish summaries of their results, and can often provide answers to specific questions, though they may charge you for the service.

Interviews

For centuries it has been the habit of scholars to do research in libraries. Often, it is better to do research with people. Ask your colleagues – they may have a story, a joke or a vital piece of information that you have forgotten. If your subject has policy implications, talk to historians, lawyers or political scientists. Is there widespread public concern? Ask a neighbour. Sometimes you can get information simply by chatting, but on other occasions you may benefit from interviewing your source formally, especially if the subject is controversial, or if your source might provide useful quotes. Quotes provide variety, authority and – with luck – a few good lines. When you interview people make sure they know why you are asking the questions, and what you plan to do with the information. You can interview by letter, over the phone or in person. It is more difficult to keep a conversation going over the phone, so have a list of questions written out before you start. If you intend to tape the call, ask your interviewee's permission first. Whatever form your interview takes, make sure that you have a permanent record, on paper and on tape, of what is being said. Check whether your interviewee minds being quoted directly, and offer to show them the finished piece. Ask the questions you think the public would want answered, and encourage your interviewee to address the public rather than you – ask, for instance, 'how would you explain this treatment to a patient?' or 'how would you explain the workings of the plant to an employee?' Ask one question at a time, and start with general matters before you go into details. Always listen to the answers, and keep an ear open for new angles. If you have a question which might antagonize your interviewee or put him or her on the defensive, ask it last. Always thank your interviewee, and ask if you may call again to check on anything that comes up when you write the article.

Not all of the interview need be reported directly. Some, particularly the information that anyone in a similar position could have told you, can be used for background. If there are parts that you want to attribute but which aren't expressed well or are too technical or full of jargon, paraphrase them and don't use quote marks – you can write 'Dr X said that . . . ' or 'Dr X told me that . . . ' rather than 'Dr X said " . . . "'. Any personal or controversial opinions should be quoted directly and word for word, although you can delete any words or phrases that aren't essential, providing that doing so doesn't change the sense.

When interviewing, be pleasant and courteous at all times – to support staff as well as to your source. Bullying people will achieve nothing, particularly since as a working scientist you cannot claim to have the might of the press barons behind you. No one is obliged to talk to you or share information, but the advantage that the working scientist has over professional journalists is that the scientist has contacts within science who may be more willing to co-operate and happier to talk than they would be with a journalist.

Doing the right thing

When you have completed your research and written your article or talk, you will need to go back to your notes about your sources to see how you stand with regard to copyright. You will need permission to publish any material – pictures or text – which you reproduce from other published works. Short quotes can sometimes be included without permission, providing they are acknowledged, but longer ones and pictures are the property of their author or first publisher, and you should ask before you publish them; sometimes you will be required to pay a fee. Unless you are planning to include a whole chapter of another book in your own, or to use a number of pictures from the same book, most publishers are happy to give permission providing you acknowledge them. It is nonetheless a necessary courtesy to ask first.

It is difficult to judge what is and isn't copyright material. Facts are not copyright, but forms of words are. Giving the source – 'as Dr Brook wrote', – may satisfy Dr Brook, but not his publisher. On the other hand, paraphrasing Dr Brook's words without acknowledging him might constitute plagiarism. Again, it's difficult to judge: many journalists work to the rule that reading one source is plagiarism, while reading two is research. If in doubt, talk to your editor or to colleagues, and think how you would feel if someone else presented your work in that way. It's also worth remembering that if you find you are still reproducing passages from other people's work, you probably haven't yet given enough thought to your own.

The problem about giving sources and references in popular work is that most people don't want any interruptions to the flow of the story. Space is limited, references can take up a lot of it. If a source is quoted it is necessary to give a reference on that page or at the end of the chapter. Be careful not to give too much prominence to the reference, and not enough to the information itself. 'The Department of Trade yesterday published a report entitled "Exporting Expertise" which reveals . . . ' still hasn't got to the point, while you could postpone saying 'as the Department of Trade's report "Exporting Expertise" revealed today' until after you had caught the reader's attention. If the name of the source might have an impact, put that at the beginning: 'Einstein once wrote . . . '.

As well as acknowledging publishing conventions, it is also worth checking whether your institution has any restrictions on publishing or broadcasting for commercial or security reasons. Going public is not a way of getting round the rigorous publication requirements of the scientific community, and you do yourself no favours by avoiding the peer review system when you shouldn't. Before you broadcast or rush into print, check with colleagues that you have their support.

Recommended reading

Sharon M. Friedman, Sharon Dunwoody and Carol L. Rogers, 1986, *Scientists as Journalists: Reporting science as News* (New York: Free Press). The first two chapters deal with 'The Scientist as Source' and 'The Journalist's World', and the third describes the role of information officers in the middle. A list of books and articles on sources is provided on pages 305–307.

Bruce L. Felknor, 1988, *How to Look Things Up and Find Things Out* (New York: Quill). Felknor charts a path through the jungle of reference books, manuals, periodicals, catalogues and computer databases in public and academic libraries. Includes chapters on science, medicine and technology.

Michael F. Flint, 1990, *User's Guide to Copyright,* third edition (Sevenoaks: Butterworth).

II. Writing

Writing

The American novelist Sinclair Lewis was once invited to launch a university course on creative authorship. When the class assembled, he stood up and said: 'Well, now, to begin with, how many of you actually want to write? Please put up your hands.' Every student raised a hand. Lewis cast an eye slowly over his audience, and then snapped: 'In that case, why the hell aren't you at home writing?'

The challenge is a serious one. Can anybody or anything teach you to write about science effectively? Some people think that writing is an art, and that only a talented few can ever write well. Others believe writing to be a craft, which develops with thought and practice. It is probably both an art and a craft – but no artist ever produced anything worthwhile without first learning practical craft skills. Here we look at the practicalities of writing and publishing.

Finding the right approach to the subject of your writing, and the right form in which to present it, is essentially the art of hitting the right target. 'The difficulty', wrote Robert Louis Stevenson, 'is not to affect your reader, but to affect him precisely as you wish.' Good authorial intentions will be of little consolation to the reader who realizes that you were never too sure of where you were heading before you got there, and that once you arrived you weren't quite sure why you'd come. A clear aim is vital to any communication project. Ask yourself these four questions:

Why should anyone be interested in what I've got to say?

A satisfactory answer to this question will tell you which aspects of your story to emphasize, and how you can relate it to your readers' interests and concerns.

Why is my subject important?

Your story may be interesting, but your readers will want to know why it is important. Thinking about the role and consequences of your subject in its widest sense can help you find its importance and guide you towards a specific readership.

Why am I doing this now?

Timing is all important, and the novelty of your subject may not be enough to carry you through media obsessed with the instantaneous production of disposable material. If your work should be presented next year, file it away and try again then. If it should have been presented last year or even last week, you won't make it into print.

Why should anyone trust me?

Decide at the outset whether you are giving information which is non-controversial, or whether the point you want to make could reasonably be challenged, in which case you'll have to convince your readers that you are right. If you don't have enough evidence of the kind that can be presented briefly and simply, and if you make extravagant claims in sensational language, no one will believe you. If you can construct a reasoned, moderate and objective argument the public will trust you and enjoy following your reasoning.

These four questions provide you with basic but essential guidelines, and should be kept in mind throughout the writing process. Ask them before you start, whenever you get stuck or side-tracked, and – most importantly – when you think you've finished.

Much advice has been offered on how you should actually set about the task of writing. 'Set aside a certain time and write every day', some say. 'What you need is application – the application of the seat of the pants to the seat of the chair', claim others. However, general advice rarely solves particular problems, and it is down to each of us to find the circumstances that suit us best. For some, a mechanical, routine process works best: the deadline, filing system and clear structure concentrate the mind and free the pen. For others, making notes, preparing an outline and setting targets don't seem to help. The medical researcher and award-winning writer Lewis Thomas remembers being asked to write a column for the *New England Journal of Medicine*:

> I had not written anything for fun since medical school and a couple of years thereafter, except for occasional light verse and once in a while a serious but not very clear or very good poem. Good bad verse was what I was pretty good at. The only other writing I'd done was scientific papers, around two hundred of them, composed in the relentlessly flat style required for absolutely unambiguity in every word, hideous language as I read it today. The chance to break free of that kind of prose, and to try the essay form, raised my spirits, but at the same time worried me. I tried outlining some ideas for essays, making lists of items I'd like to cover in each piece, organising my thoughts in orderly sequence, and wrote several dreadful essays which I could not bring myself to reread, and decided to give up being orderly. I changed the method to no method at all, picked

> out some suitable times late at night, usually on the weekend two days after I'd already passed the deadline, and wrote without outline or planning in advance, as fast as I could. This worked better, or at least was more fun, and I was able to get started.[1]

Thomas worked to a schedule others would find terrifying, and worked to it successfully. His collections of essays, *The Lives of a Cell* and *The Medusa and the Snail*, have been critical and popular successes.

The process of writing is difficult enough without placing unnecessary impediments in your way. Make yourself as comfortable as possible: the physical process of writing should be done wherever, whenever and in whichever way suits you best. Don't chain yourself to a typewriter during the small hours if you'd feel better writing in pencil before breakfast. Be thoroughly familiar with your material before you start. If you get stuck, put the work away for a while and think about other things. Sometimes talking about the subject with others can get you started again. There are a number of different stages to the writing process, and you may find that you are tired of plotting the adventure story in the middle of your piece, but would be quite happy to spend some time polishing your introduction. If you get stuck with the beginning write the middle, and then see what sort of beginning your middle needs. You may then find that you have to rewrite the middle, but at least you'll have a beginning. Try not to stop when you are stuck, but when you are in full swing. Take a break in mid-paragraph, or even in mid-sentence. You'll have an easier start next time you sit down to write.

Getting stuck on

> I was once told that the surest aid to writing was a piece of cobbler's wax on my chair. I certainly believe in the cobbler's wax much more than in inspiration.
>
> Anthony Trollope, novelist

Sooner or later (and generally sooner) in their careers, all writers encounter problems in getting started – you are sitting in front of a blank sheet of paper, and a kind of mental and physical paralysis sets in. You decide that the surroundings are not quite right; the weather is too good to work; that you need to find a few more references; suddenly you're thirsty . . . You are suffering from writer's block, and the only way to deal with this is . . . to start to write. That may sound unhelpful, but writing something is better than not writing at all. Start by writing a plan, and if a plan is not forthcoming write lists of everything you want to say, order the lists, and then see if a plan emerges. If it doesn't, you may not be ready to start writing, and would benefit from a little more preparation.

You may find that you can't get a single word down on paper, but could talk with friends for hours about the subject of your writing. Speaking is, for most people, much easier than writing. If that's how it is for you, then write down what you would say. To write exactly as you speak would produce something quite unreadable for anyone who didn't know you – a little like a rambling letter to a close friend – but once you have the words on paper you will find it much easier to organize them than you would if they were still in your mind.

Speech is a valuable basis for writing because when you talk you are conscious of the people you are talking with. Writing is in many respects a private substitute for conversation, and it is important to make it bear as much as possible on the interests and experiences of readers. Since you will not be around to help each reader make sense of your story, you must be accurate and precise when you write. These qualities are especially important when you are dealing with technical matters or with matters which if misunderstood could create anxiety or harm. The usual, basic defence of 'good English' is that it helps communication, and thereby mutual understanding. This is a crude truth, but it is also the most valid and pressing argument in favour of the correct use of language. You can sometimes get by with ambiguous and imprecise language – everyone figured out what an early manual on child-rearing meant when it gave the immortal instruction 'If the baby does not thrive on raw milk, boil it' – but 'figuring out' is precisely what your readers should not have to do. They are busy people, and don't have time to decipher your personal code. Your readers – however intelligent or knowledgeable – will skip parts or give up on an article if it is too obtuse, so that a brilliant ending won't compensate for a boring middle. Don't bank on their reading it more than once, and don't expect them not to notice ambiguities or contradictions. Some of your fiercest critics will be non-experts.

Reference

1 Lewis Thomas, 1985, *The Youngest Science: Notes of a Medicine Watcher* (Oxford University Press), pp. 242–43.

2.1 Outlets and Opportunities

The print media are by far the most accessible of those available to the non-professional. The recent explosion in the newspaper and magazine industry has created more space to be filled, and has greatly increased opportunities for freelance or occasional writers to get into print.

From the consumer's point of view, the written word is a good medium for science: readers can stop for breath or retrace their steps if they lose their way. Printed material is also easy to store away for future reference. Your efforts in writing will be repaid by a potentially huge audience and, if you write for magazines or books, your work will be preserved for many years. Newspapers are more ephemeral – many journalists have been depressed by the thought that their work represents a temporary stage on a massive conveyor belt which moves from pulp to pulp.

Because you are a specialist in your field, you may well be invited to write. It is always flattering to be given the opportunity to express yourself in print – be it for a national newspaper or the in-house newsletter – and it is easy to accept. But before you do so, think about whether you are comfortable with the attitude and style of the publication, and whether you have time to do the job properly. Think about whether you are qualified to write that particular article. If you are sure you want to accept the invitation, ask plenty of questions before you start. How long should the article be? Will you be allowed illustrations? Who is going to read it? Will you be paid?

Most of us who have ever had the urge to put pen to paper can't afford to sit around waiting for an invitation. Fortunately, the print media are open to all-comers – anyone can write. The difficult part is getting published.

There are four main forms of science writing: news, newspaper and magazine features, books, and book reviews.

News

Writing news stories is a specialized art and operates to very tight deadlines. Because journalists have little time to spare for polishing their work they have to get it right first time, so rules and practice are strictly observed. A glance at the news section of science magazines or at the science columns in newspapers will show you that most science news is written by professional writers on the publication's staff. Even if you produce your news item at lightning speed and dispatch it by express to the news desk, the chances are that, if the editor wants your story, it will be used as a source rather than as a finished piece.

Here is an example of how a science story might be turned into a news story. Your abstract reads:

PROBE RESULTS MODELLED

> Recent computer and practical simulations have been performed which suggest a possible description of the nature of the readily observed phenomenon, the so-called Red Spot on the planet Jupiter. Based on probe observations of the gaseous nature of Jupiter's surface and the well-known banded appearance of its atmosphere, the computer simulation

and the experiment, which was carried out in a circular vessel with strategically placed valves which control the flow of water, both resulted in patterns not dissimilar to those observed on the surface of Jupiter, including the spot. Jupiter rotates completely during a ten-hour period, which causes movement of its atmospheric gases in a complex and dynamic, alternately directioned system of bands, between which develop regions of shear, resulting in the formation of vortices which coalesce to form one larger one, the Red Spot.

The news story might read:

JUPITER SPOTTED

The planet Jupiter's surface is covered by flowing bands of gas, which give the planet its striped appearance. The bands were discovered by the Voyager space probe. This information has helped scientists in California to suggest a 'whirlpool' explanation for Jupiter's mysterious Red Spot.

The scientists have tried to imitate the flowing gases in their laboratory. They used a water tank fitted with valves, pumping water through the valves to produce the flow patterns seen on Jupiter. The tank was spun round to imitate Jupiter's rapid rotation. This rotation caused bands to form in the water, which formed a striped pattern like that on Jupiter. Each stripe moves in the opposite direction to the stripes on either side of it. In the tank whirlpools formed where the bands rubbed against each other, and some of the whirlpools merged to form a larger one.

The scientists think that Jupiter's Red Spot could have been formed in the same way as the whirlpool in the tank.

The first of these is not a news story – the second could be. The second version has been rewritten and edited so that it has a clear structure, is written in short sentences and uses less jargon. The editor was not happy with phrases such as 'simulations have been performed', and wanted some people in the story, so he rang you to find out about the California group. Voyager had been in the news, so it was used to provide a signpost for readers in the first paragraph. The first version of this story is only useful as a source of information, so you might as well have rung the paper's science correspondent in the first place and saved yourself the trouble of writing it down. You won't be paid for it. That's the bad news.

The good news is that you can provide a useful service by keeping science correspondents informed of events as they happen, and you can contribute to science news by writing about events before they happen. This may encourage the correspondent to attend the event, if it is a meeting or a lecture, or to prepare some background if some important research is about to be published. Be selective though: if you bombard your local paper with a torrent of uninteresting stories the staff may get into the habit of ignoring your suggestions.

Some newspapers have special weekly science features and all of them carry science stories in the news sections, but there you will be competing for space with politics, finance, international news and arts. The *New York Times*, for example, carries the feature 'Science Times' every Tuesday, which is syndicated across the United States. Although they have twelve or so pages to fill every week, the staff of Science Times write all the articles themselves, and find their stories by keeping up with the specialist sources and through personal contacts in the field. The *New York Times* does not publish unsolicited science articles by freelance writers, but it is an excellent place to look if you are trying to get a sense of what good science journalism in newspapers should be like.

If you look at the articles in the *New York Times*, you will see that the same authors are writing every week. This is reliable evidence that the articles are all written by the staff. A few newspapers, for example the British quality papers, publish stories in their science sections by a variety of authors, and so are more likely to publish freelance work. Typical of these is the British newspaper the *Independent*, which runs a page or so of science news and features once a week. Whatever publication you choose, be thoroughly familiar with it before you start to write. The most common single reason why editors reject articles is because their authors have not thought carefully enough about what the readers will find interesting. A subject which is of passionate interest only to you is probably not good material for a national newspaper.

Caring and sharing

> Many scientists assume that what professional scientists care about will also fascinate readers. Generally, though, this is not the case.
>
> Daniel Goleman, science reporter, *New York Times*

If you have no luck with the national papers on your first attempt, try once more during a slack period, such as the parliamentary recess or just after Christmas. Local newspapers are a useful outlet; they cater for a readership that may well know about your place of work and be curious about what goes on behind its closed doors. If your local paper doesn't usually feature science, you might encourage them by submitting a story or inviting a reporter to visit you when you have some news to share.

Compared to academic publishing, newspapers and magazines publish very quickly. However, unless your article is planned to coincide with a particular event it may not appear for some time. You are competing for space with stories which change from day to day, while the contents of your story may not change for years. This gives science stories an advantage in that they can be brought out to fill a space long after other stories have been discarded. Editors will be more receptive to your article if it is directly relevant to another news item, so it might be worth putting it aside until some event makes it noteworthy – irradiated food goes on sale, funding is withdrawn from space research.

When your article is accepted it will be edited. Recognizing an editor's responsibilities can help you understand what happens to your article after you have sent it to the science correspondent or features editor. The role of journalists and newspaper editors is to report and comment, not to protect or promote. You may want to inform, while the editor wants to entertain. Editors are constrained by time, space, political commitments, sales and advertising. Very rarely is an article the exact size for the space available, although reducing or enlarging pictures can sometimes provide the necessary adjustment. Even if you are shown the edited version before publication (which very rarely happens), you should expect some surprises in the published version – there may be paragraphs missing and it is bound to bear a different title. Comparing the published version with the version you submitted can give you an insight into the editorial process which will be valuable next time you come to write.

News of your work may appear in a newspaper as the result of a press release. Many companies, laboratories and research institutes have information officers whose job it is to prepare press releases. However, since every large institution issues a superfluity of them, many journalists treat press releases as junk mail and ignore them. While some reporters will reject all press releases, others will be selective and will know a piece of genuine news when they see it. They can also detect hype, and once they start getting rubbish from a particular source they will soon stop opening the envelopes.

It pays to liaise with your information office, to supply it with newsworthy stories, and to make sure that it is sending interesting press releases to the appropriate publications. A good press release will be written in the style of a news story, so they are well worth studying if you plan to write news.

The news story is one of the most disciplined forms of writing. The immediacy of the introductory paragraph, the carefully controlled narrative structure, the strict accuracy, relevance and topicality of information, the economy of language, and the neutrality of the editorial viewpoint – these features give the news story as distinctive a style as the technical paper or laboratory report. Soon after the Second World War, Earl Johnson, general manager of the news service United Press International, issued a memorandum which read: 'much of the news these days is so important that it deserves to be understood by the widest possible audience. Let's have more periods and fewer complex words. Watch for that lead sentence. Keep it short and simple. Then let the lead set the pace for the whole piece.' That advice soon spread to all the news media and even the 'serious' papers began to publish stories that were easier to read and easier to understand. When Armstrong and Aldrin landed on the Moon in 1969, the *New York Times* opened with the simplest possible lead: 'Houston, July 20 – Men landed on the Moon today.'

The news story is essentially a self-contained factual report, but unlike a fictional story, which unfolds gradually until a complete pattern emerges, the news story begins by encapsulating the main features, and then summarizes the circumstances that surround them. In other words, it begins with the conclusion and then states the main findings that lead to that conclusion. Here's a good example taken from the *Independent*:

ASTRONOMY THREATENED BY 'RADIO POLLUTION'

The future of astronomy is threatened by the ever-increasing number of radio transmissions, the Astronomer Royal warns. Sir Francis Graham-Smith says satellites, television, radar and cellular telephones are swamping frequencies used by astronomers.

'With the multiplicity of commercial and military satellite systems, there may soon be nowhere on Earth where radio astronomy can be carried out,' he says.

Radio astronomy enables us to 'see' almost as far back as the original Big Bang. But Sir Francis, whose book *Pathways to the Universe*, to be launched next Monday, attempts to make astronomy accessible to everyone, is worried that civilisation is rapidly threatening to close our 'windows' on space.

His warning about 'radio pollution' follows concern about the effects of 'light pollution' in and around cities, which makes it impossible to view the stars from optical observatories nearby.

Radio telescopes record the extremely faint radio wavelength emissions from distant stars and galaxies. The Science and Engineering Research Council has just announced a multi-million pound investment in a large new telescope near Cambridge.

Editors cut news stories from the end, so put the most important information in the first paragraph, and follow that with the rest of the information in decreasing order of importance. The radio astronomy story you have just read illustrates this very well. The story has to stand alone and make sense without the last paragraph, or even without the last two or three paragraphs. The opening statement must be immediate, positive and active. It should go straight to the point. For example, do not write: 'According to a report published today, British children are healthier than ever before' – write: 'British children are healthier than ever before, according to a report published today.' The statement should be positive because news records change rather than the absence of change. Do not write 'No one was seriously injured when a fire broke out in a laboratory at British Chemical Research last night' – that suggests that nothing happened. 'Scientists escaped injury last night when a fire broke out in a laboratory . . .' would be better. And finally, use the active rather than the passive voice. 'Oxford physicist Dr Joan Simmons has won an international science prize' is more interesting than 'An international science prize has been won by an Oxford physicist, Dr Joan Simmons'. The first version makes better news because it is about a person, while the second version is only about a prize.

Editors have a great deal of influence over how an article appears in print. Unlike scientists, who expect to see what they have written appear exactly as they have written it, journalists are surprised if this happens. A whole team sets to work on the piece once it has left the writer's hands; it gets copy-edited, headlined and positioned, all without consultation with the writer. Editors, like physicists, have to contend with problems of space and time. For the editor, both space and time are in short supply. Warren Leary, a science writer for the Associated Press wire service, explains: 'we have to look at our audience, our time constraints and our space restraints. I don't frequently have 900 or 1200 words to explain something in a story. I might have 400 or 500 words.' The editor's job comes with a sharp knife and a blunt pencil: the first to cut, the second to delete. The best way to prepare yourself for the operation is to make it unnecessary for the editor to rewrite your news story and to make it easy to chop.

Although it is unlikely that you will have the opportunity to write a great many news stories, the lessons you will learn by forming concise, dramatic, immediate and compelling stories are useful when you come to write features as well as books.

Features

A feature is the ideal format for science stories as it offers room for a structured argument, detail, history and pictures. The publication schedule is more flexible than for news, and you may even have a chance to check the editor's alterations to your work. There are countless outlets for freelance feature articles about science, though many of them, particularly in the UK, are rather difficult to spot. Articles on scientific subjects crop up in magazines as diverse as *Amateur Winemaking* and *Vogue*, in the local and national newspapers, as well as in staff magazines and newsletters. Try current affairs magazines – the *New Statesman and Society* or the *Economist* – or magazines aimed at women or people with particular hobbies. Magazines about science are more popular in the United States, and many of these are read the world over. Try *Discover*, *The Sciences* or *Popular Science*. There is a big market for popular science magazines in Europe, and don't be shy about submitting an article to a magazine which is published in a language other than English. The staff of these magazines will shame you with their facility for languages, and will always be able to translate what you send, so look out for magazines when you travel. France produces the popular *Science et vie*, and the excellent *Scienza e Dossier* is published in Italian. A useful guide to available outlets is *The Writers' and Artists' Yearbook*, which gives advice on how to prepare your work, and lists magazine and newspaper titles and addresses along with information about payment, illustrations and interests. The American equivalent is *Writer's Market: Where to sell what you write*. Since publishing – particularly magazine publishing – is an unpredictable and ever-changing business, these writers' guides are published annually, so be sure that your copy is up to date.

If you are writing a feature article, and have found a likely home for it, read the publication thoroughly to get an idea of the readership and the house style – the level of the writing, the length of the articles, readers' letters, advertisements and cover price all offer clues. Think about the style of the writing, the complexity of the information presented, and whether the publication has any affiliations or particular interests.

Cut!

If a reporter asks how long a feature story should be, the editor may reply 'as long as you keep it interesting.'

Daniel R. Williamson, professor of journalism

If your article is long, try sending the editor an outline or summary. He or she is more likely to read something short, and if it's not suitable you will have saved time and postage. However, a subsequent invitation to submit the full version is not the same as a commitment to publish.

Although feature articles do not have the same urgency or as short a life time as a news story, they benefit from being timely. It is often possible to turn a piece of science writing into a topical feature by linking it to an anniversary.

One hundred years ago Dr Felix Hoffman developed the aspirin, but we still know remarkably little about how this drug works.

When jazz-players and cotton-pickers in New Orleans reach for their bottles of bourbon this year, few of them will spare a thought for Mrs Eliza Craig. But they should, for Mrs Craig was the first person ever to blend bourbon, 200 years ago. Bourbon presents some fascinating problems for the scientist today.

People are just as interested in the shape of things to come as they are in the way things are and used to be. You don't have to leap into the science fiction world of atomic monsters and faster-than-light spaceships to write about innovations and developments that seem likely in the near future. Examples might include biological computers, tidal power stations, offshore cities and artificial tooth enamel. By relating present-day science and technology to possible futures, you will discover an intriguing way to provide a setting for what might otherwise seem abstract pieces of scientific and technological research. Within bounds, 'futurology' is a good way to deal with the 'so what?' response some readers may otherwise have to your writing.

Features generally pay better than newspaper articles, although some magazines pay newspaper rates. While they vary wildly, examples of the payments you might expect for your work are £160 per 1000 words in a newspaper or a news magazine – this figure is from *New Scientist* – and anything from $50 to $500 for a feature article – this is offered by the magazine *Astronomy*. Again, the message is clear: if money is your motive, don't bother.

Books

If you write a newspaper or magazine article, your responsibility ends the moment you post it. A book, on the other hand, will take years to write and will occupy your spare time for months after you think you've finished. Books are an enormous investment in time and effort, and unless they are unusually successful they reach far fewer people than a feature article. They do, however, last far longer and will be available in libraries – if they are published. Knowing a lot about a subject doesn't necessarily qualify you as an author, and unless you're like Molière's Monsieur Jourdain, who discovers he's been talking prose all his life without knowing it, writing a book can be a terrible ordeal. Charles Darwin took nearly a year to write his first book, an account of his voyage around the world on HMS Beagle. 'I had no idea', he wrote, 'of the trouble which trying to write common English could cost one'. Nonetheless, the book was a popular and lasting success.

Think before you start about your other responsibilities, the equipment and help you will need, and your qualifications for the job. Look at the project from a publisher's point of view: the likely market, the format, the pictures, the size of the book and its price, and the range of similar publications.

You might think about collaborating with a professional writer. This can be enormously helpful when you are writing for the general public, who expect far higher standards of their reading material than scientists apparently do. A non-expert collaborator can also help by acting as translator if you are being too technical. Novice authors, particularly scientists, often receive valuable help from their publisher, so it is important to make contact before you start, and to keep in touch while you are writing. Since every book is different it is difficult to make general recommendations about how to go about writing one, but it is your editor's job to help you, so don't be shy about asking for advice whenever you need it.

When you are choosing a publisher, the first question to ask is whether the company you have in mind actually publishes books in your field. If in doubt, go to a bookshop and see who publishes the books which are like yours in subject matter, length and price. Try to assess how effective the publishers' promotion and publicity departments are. Do you receive booklists and catalogues from them; are they represented at conferences; do they advertise regularly? What about overseas sales and translations? Talk to colleagues, talk to booksellers, and read science journals and magazines – not only the reviews but also the publishers' advertisements.

There are no ideal publishers: small companies give you personal attention but haphazard distribution; local publishers are conveniently near to hand but may as a result rely on you to do the work which they ought to be doing themselves. Choosing publishers can take some time, but it's well worth the effort – otherwise you may see several years of toil and trouble reproduced in an over-priced, under-resourced, and obscure book.

It is difficult to know, before you start, whether you actually have a book's worth of words to write, even if you are planning a comprehensive catalogue of the world's insects. It is also difficult to devote a year's spare time to writing if you don't have the reassurance of a publisher's contract which tells you that if your book meets the specified standards it will be published. Fortunately you can solve both of these problems by writing a proposal for prospective publishers. Preparing the proposal will tell you whether you've actually got anything to say, and when publishers read it they will tell you whether or not the book is worth publishing.

Your proposal should be a short document – two or three sheets of paper, no more – which should include a brief CV, a description of the book, the reasons why you think it should be written (including an analysis of its likely market), and details of your proposed schedule.

Include in your CV a list of your publications (popular and academic) and describe any relevant experience. If you have published any books before, the commissioning editor may ask to see sales information or reviews, so have these to hand, though it is probably not worth including them in the initial proposal.

Next, you should write a description of the book. First comes its title. The title is a vital component of any book, and merits some thought. It should tell the reader exactly what the book is about, but without giving everything away. The title is a big clue to the tone of the book: if it is a joke it had better not be the only one in the book. There are also practical considerations: it must fit onto the spine of the book. If the title is long, it will be abbreviated in a number of situations, and this may make the book difficult for readers to find.

Then you should write 300-500 words about the contents. List the chapter headings, and briefly describe each chapter. The third section of the proposal is one of the most important to publishers, yet the most neglected by authors. Anyone thinking of offering you a contract will ask the following questions, so answer them in your proposal.

Why should the book be written?
In what ways is it original?
Why are you the right person to write it?
What other books already exist on the subject? (A brief analysis of these can be very helpful.)
Who will buy it?

Editors generally find that the author's description of the likely market for their book is the most unsatisfactory part of any proposal, so make sure that you are clear about who you are writing for, and that you are realistic about the range of readers who may find your work interesting.

'The interested general reader'

One big warning: please don't put 'The interested general reader'. This legendary character may exist, but the term has become something of a standing joke among publishers. Also don't delude yourself that a book on a very specific topic will have a wide audience. There are of course always surprises in publishing, but on the whole *Fish Farming in Southern Sudan* is not going to be a successor to *Life on Earth*.

Romesh Vaitilingam
Senior Commissioning Editor, Basil Blackwell

If your book is a textbook, at what level will it be used and on what courses? Are there any particular parts of the world where the topic will appeal? Finally, any thoughts on sales to book clubs and professional associations are always welcome.

When you have provided this information, it remains to provide a realistic date for the submission of the final manuscript (you would be wise to overestimate) and to describe the physical appearance of the book – format, illustrations and length. It is difficult to estimate length in advance, so find a book which is about the size you expect yours to be and work out the number of words in it. If you have started writing, you could offer to show the publisher your efforts to date.

Having discovered one or more suitable publishers, your task is now to contact the commissioning editor. This you should do by letter and by name. Don't telephone the editor before submitting a proposal, and don't send in a 'Dear Sir or Madam' letter: it will look as if it has already done the circuit. Your proposal and a short covering letter going straight to the point are all that is needed.

Once your proposal is accepted you will make a formal agreement with your editor. The publisher must meet three objectives: to fix a realistic price, to reach a sensible gross margin, and to maintain the editorial strategy and reputation of the publisher. When a contract is offered to you, you will see that it specifies subject matter, format, length, schedule, and how you will be paid. Contracts are without exception long and tiresome documents full of legal jargon. As with any contract, remember to read the small print; since it will in all likelihood all be in small print, this means read it from beginning to end.

The clauses that are most likely to have a direct impact upon you are those about copyright and author's royalties. Read these carefully and discuss them with experienced colleagues. Royalties – that is, how much you may be expected to earn as a percentage of net receipts or sales – are in principle negotiable. Although it is unlikely that one publisher will offer substantially more in royalties than another, it is well worth discussing an advance, returns on foreign sales, and even such eventualities as television or newspaper serialization. How about the number of complimentary copies? Translation, television and anthology rights? If there are any illustrations, who will have to secure reproduction rights (often a tiresome business), and will you be asked to bear the cost of that yourself?

You should remember that when all talk of 'enhancing public understanding' and 'contributing to scholarship' is over, publishing is a business. Your editor is partly concerned with promoting scholarship and partly with promoting sales, and this can sometimes mean that compromises must be made. Establish a friendly relationship with your editor, who can be a great help to you throughout the writing and publishing process. In return you should acknowledge that all of your publisher's staff are professionals working to strict schedules, and you should do your best to stick to your side of the deal. This means you should present to your publisher on the date specified in the contract a manuscript which follows the specifications of length, style and content given in the contract. Don't think that you can submit a badly prepared manuscript and then polish it up in proof – even minor proof corrections can result in changes to page layout and numbering, and to the contents list and index. Proof corrections are expensive, and the cost may be deducted from your royalties. The time involved will delay the publication of your book, which may have been timed to coincide with a particular event (a space launch or Christmas). This will affect sales and jeopardize the publisher's investment. You can avoid all these problems by being completely sure of your manuscript before you submit it. If you do get behind schedule, let your editor know as soon as you can.

Book reviews

Reviewing is almost as old an occupation as writing. In 411BC the Greek satirist Aristophanes called the great tragedian Euripides 'a cliché anthologist . . . and maker of ragamuffin manikins'. Two thousand years later it is still the case that one of the hazards of writing a book is that sooner or later someone will write a review of it. One of the hazards of having colleagues who write books is that sooner or later you are bound to be asked to write a review. Book reviews are a specialized form of writing, yet they are extremely versatile – they can be a vehicle for vanity or vengeance, a nightmare or a joy. Book reviews also provide a valuable service to book buyers, and are an excellent way of bringing science books to wider public notice. Reviewing also carries considerable responsibility; while your reviews may never achieve the power of the make-or-break pronouncements of the drama critic of the *New York Times*, many people will base their judgement of the book on what you write.

Some wit once said that the proper way to review a book was first to write the review, then to read the review, and then if the review sounded promising to read the book. It is easy and tempting to give a book a cursory glance, absorb the blurb, rush through a few pages from the introduction and conclusion, and then set down some magisterial words of praise or damnation. However, the first duty of the book reviewer is to read the book.

The second temptation is to offer an immediate response, particularly if it is vitriolic. Unless you are asked to deal with a book that has no redeeming features at all, a wholly negative review will not be published. At the other extreme are reviews which consist of unreserved praise, but think twice if this is your reaction to a book: what is timeless one year is outdated the next, masterpieces are quickly replaced by better works, and unique contributions rarely remain so for long.

Overkill

> I have long felt that any reviewer who expresses rage and loathing for [a book] is preposterous. He or she is like a person who has just put on full armour and attacked a hot fudge sundae or banana split.
>
> Kurt Vonnegut Jr, novelist

Writing a balanced review which avoids these excesses takes time. You may need to read the book more than once. Make notes as you go along, and then put the whole lot away – book and notes – and let your thoughts settle for a few days. You will then be able to distinguish between lasting impressions and trivial quibbles which aren't worth recording.

Book reviewing is not an exact science, but there are certain features which any competent review must include, whatever its length or destination. Samuel Johnson used to ask three questions of any book, and so should you:

What did the writer try to do?
How well did he or she do it?
Was it worth doing in the first place?

Finally, a review is not the place to promote your own opinions about the book's subject, or any other. There is no need to talk at length about the progress of twentieth-century physics in reviewing a book of essays on space travel; nor should you use the first paragraph of a review of *Genetics Today* to state your own views on cloning. Book review editors like reviews which are informative and entertaining, short and snappy, and which present a fair and accurate summary of a book along with comments and opinions about it. They like strong opening 'hooks' to catch readers, and dislike phrases such as 'This book is about . . .' and 'In this book . . .' and banal endings such as 'on the whole, all things considered, I believe . . .', or which 'thoroughly recommend this book to every scientist.' They dislike extended summaries, lists of chapter headings and long quotations. They will delete your remarks about the size and price of the book on the assumption that readers can make up their own minds from the publisher's information which will be printed with your review.

Many review editors have a list of tried and tested reviewers and only accept commissioned work. However, there are plenty of outlets for your own work and you don't need to wait until you are asked. Take the initiative by writing to the editor of a magazine or newspaper suggesting that you review a particular book and mentioning a few reasons why. Local newspapers, college papers, quarterlies and newsletters should all be responsive. While science books seldom get star treatment, they are considered by review editors and, after some experience of reviewing for a small-circulation paper, you could send samples of your work to a publication with a broader readership.

You will be able to see the style of review favoured by different publications and you should try to adapt your writing accordingly. Reviews in *Nature*, *The Sciences* and *New Scientist* are almost always excellent and can give you some examples of the proper use of quotation and summary, and of fairness and judgement in reviewing. They might even have received Dr Johnson's blessing.

Recommended reading

James Aitchison, 1988, *Writing for the Press* (London: Hutchinson Education). A short, lively guide to news reporting, which includes chapters on news sources, news values, prose style, using quotations and interviewing.

Warren Burkett, 1986, *News Reporting: Science, Medicine and High Technology* (Ames Iowa: Iowa State University Press). A comprehensive account which includes chapters on choosing science news, gathering information, how to avoid distortion, and ethical issues in science writing. Includes a useful bibliography.

Maurice Dunleavy, 1988, *Feature Writing* (Geelong, Victoria, Australia: Deakin University Press). A concise account of 'how to write a feature article in one easy lesson', along with a further lesson on advanced techniques, and three practical chapters on interviewing.

Brendan Hennessy, 1989, *Writing Feature Articles* (Oxford: Heinemann Professional Publishing). Full of helpful advice on style, substance and structure.

Sharon M. Friedman, Sharon Dunwoody and Carol L. Rogers, 1986, *Scientists and Journalists: Reporting Science as News* (New York: Free Press). An insightful study of the relationship between science and the mass media based on analyses from both points of view. Lots of practical advice, and annotated bibliographies.

2.2 Style and Substance

Most scientists write only for specialist journals. To do this they have to learn and stick unflinchingly to a set of rules and conventions that authors in any other field would find unreasonably and painfully constricting: they must obey regulations which prescribe not only the style in which they must write, but also the form their paper must take. The standard scientific paper follows an established pattern: there is an abstract, which presents the entire contents of the paper and thereby saves the reader from having to read it; an introduction, which restates the abstract and presents the aim and scope of the paper; a 'materials and methods' section which restates the abstract and the introduction, sets out the overall experimental design or theoretical approach and the premisses and assumptions made in the design, and justifies the choice of methods; the 'results' section, which restates the abstract and introduction and presents results which justify the choice of methods; the 'discussion' section, which begins 'As stated in the Introduction . . .', justifies the choice of methods, defends the results and then apologises for them; a summary which restates the abstract, introduction, method and results; and then sections for acknowledgements, where one thanks one's patron and the journal editor so that the next piece of research will get funded and the results will get published; appendices, where one can make another attempt at explaining that part that didn't come out right in the text; and references, which are really the acknowledgements . . .

The scientific paper allows no room for the author to express the reality of scientific research. Scientists must distil from a mass of data and experience only a few specific things. They must convey the results of an enquiry and not trouble the reader with an account of its history – they record the researcher's arrival at a destination and claim that the journey was smooth and direct, with no wrong turnings, detours or halts. Moreover, it is a journey undertaken without visible travellers: scientists must write in an impersonal style which obliterates any evidence of their own experiences, opinions and feelings.

Scispeak

> As things are, too much of what passes for the scientific literature is not literature at all but a way of stringing code words together in such a way that the perpetrators can enjoy the warm glow of knowing that a piece of research has been written up and given a prominent place on the library shelves throughout the world . . . . The immediate interests of readers that they should be able to read and understand are given only scant attention.
>
> John Maddox, Editor, *Nature*

Scientific papers were originally designed to carry the maximum of information in the minimum of words. When in the seventeenth century the members of the Royal Society considered their new journal, *Philosophical Transactions*, they agreed that they would use 'the language of artizan countrymen and merchants, before that of wits and scholars.' Despite the good intentions of these early scientists, three hundred years later we can see by reference to any scientific journal that their successors have not followed their lead.

Some people, reared on a diet of academic texts, find the highly formal structure easy to imitate, and so write happily in the time-honoured style. Some readers like the format because they know that if all they need is results they can dig out the appropriate section and not have to rummage around among lots of superfluous ramblings about scenery or exciting times down at the laboratory. But scientific publications have purposes other than the communication of ideas: they represent the productivity and therefore the 'value' of the research team; they establish hierarchies by the ordering of their author lines and by whom they chose to cite; and, most importantly of all, they stake their author's claim to the new knowledge they contain. They serve the needs of their authors above the needs of their readers.

In popular writing, however, the top priority is communication with the readers. While there are some lessons to be learned from writing scientific papers – clarity, conciseness and accuracy are as important in a newspaper article as they are in a contribution to *Nature* – other aspects of the scientific paper will impede effective popular communication. Instead of writing in an impersonal style, you will need to describe your personal involvement and feelings. Readers will be interested in learning about how results were obtained (*really* obtained) as well as their significance, and will want to see what's going on behind the scenes as well as on centre stage. Perhaps more important than all this, you will need to adapt to the freedom offered to you: what you write about will be limited only by your own skill and the interests of potential readers. How you write, in what style, at what length, with what examples and illustrations, and with what objectives in mind, is up to you.

DNA codes

> Just over 30 years ago *Nature* published a report of what is now recognized as one of the key discoveries of modern biology. Like most scientific papers it was written in a kind of code. Who outside the small group interested in DNA could have recognized the significance of Watson and Crick's structure from their letter alone? . . . the scientific paper is a very formalized mode of communication; it makes no attempt to put the work described in any kind of broad perspective, and deliberately eschews extrapolation and speculation. Yet for the general public the first is essential, the second desirable and the third often useful. Governments fund research and the taxpayers pay for it because they believe it will bring benefits. They are entitled to know what is just around the corner and what is looming over the horizon.
>
> Bryan Silcock, Senior Systems editor, *Sunday Times*

In popular writing an author's job involves taking sides, expressing opinions and conveying the impression you want to convey. You are writing to convince: to convince your readers that your subject is important, that the information it contains is true, that they can trust your authority, that articles about science are worth reading. You may want to expand their knowledge or understanding, or simply to brighten a rainy afternoon. You may be offering straightforward facts, or you may be expressing an opinion or preference: you will need to convince the readers that your line of thought is worth considering. You may be offering judgement: you want the readers to believe that you are right. No matter what your approach is, you are trying to elicit a particular reaction from your readers. You are trying to persuade them to perceive, to admit, to agree. You want readers to see things as you have seen them; you have designs on their reason and imagination.

So how can the language you use help you to put your case effectively? Different styles of writing suit different objectives, and it is important in successful communication to chose a style and form which are appropriate both to your readers and to the purpose of your article. If you want to write 1500 words for teenagers on the development of our understanding of superconductivity, a historical, story-telling approach would be helpful; an article in *Motorcycle Mechanics* on fuel efficiency might be more discursive, and consider how technological advances have resulted in practical changes for today's motorcyclist. It will certainly help your writing if you consider what different styles can accomplish.

A scientist's guide to style

In 1753 the French naturalist Buffon enjoyed the rare distinction – especially for a scientist – of being elected to the Académie française, the organization that keeps a watchful eye on French language and literature. Buffon gave his inaugural address on the subject of literary style, and the speech became famous for the line 'the style is the man'. But Buffon in the same speech offered another definition of style which is more useful for the would-be writer. 'Style', he said, 'is simply the order and movement one puts into one's thoughts.'

This sentence encapsulates the essence of clear, engaging writing. Order provides the logical structure which is essential to the reader who is trying to follow a story or argument. By the time your article is in print you have to let it stand on its own, and you can't be around to remind the reader of the hint you gave in paragraph seven or of how it all gets explained towards the end. The moment your story stalls or starts wandering you'll have your work cut out to keep hold of the reader's attention. If the readers think, 'Wait, I don't quite follow that' or 'Have I missed something?', you are beginning to lose them, perhaps irretrievably.

Order in a story is perhaps easier to find than order in an argument or explanation. If you find yourself writing 'and now to get back to those electrons I mentioned earlier' or 'as we will see later' then your order needs attention.

One horror enjoyed by academic publications is the footnote. Noel Coward detested footnotes, and compared the eye's trip from the text to the note to having to go downstairs to answer the doorbell in the middle of an amorous dalliance. He had a point: notes are always set in minuscule type, and are sometimes even hidden away at the end of the chapter or the back of the book, and so interrupt the flow of the text. Any order in your writing is destroyed while the reader searches for the note. Because of this inconvenience readers often ignore footnotes, so if the material in your footnotes is important, it should be in the text. If it isn't important, you shouldn't be wasting the readers' time with it.

Buffon's second quality, movement, is what carries the reader from one idea or event to the next. It is what ensures that the reader who picks up your article will see it through. Words like 'while', 'next', 'suddenly', 'first', 'later' and 'afterwards' give a sense of time, and phrases like 'as a result' and 'this led us to believe that' show a logical progression. The length of your sentences gives an impression of pace. Long flowing sentences give an impression of relaxed, easy and smooth sequence of events, while short sharp sentences speed up the action. Compare these two descriptions of the same event:

> Mission control broadcast the countdown to the spectators at the site, who watched the rocket fire and disappear behind a cloud of smoke.
>
> Mission control broadcast the countdown. The spectators held their breath: five, four, three . . . The engines fired. The rocket disappeared behind a cloud of smoke.

The second version, with its shorter phrases, more effectively communicates the excitement of the occasion. For the same reason, quoting the actual words of a participant or witness can enliven a story – people very rarely talk in long sentences.

The shock of the new

The paradox of writing about new discoveries is that while you have the advantage of the subject's intrinsic novelty value, you have the disadvantage of dealing with material which is unfamiliar to your readers. The most effective way to do this is to start from the familiar and simple and then proceed to the unfamiliar and difficult. The key to some of the best popular science writing lies in the author's ability to make connections that others miss. J.B.S. Haldane likens the production of hot gas in a bomb to that of steam in a kettle, and the changes which occur in a bird each year to those which take place in people once in a lifetime at puberty. Analogies such as these provide a familiar setting which makes new ideas less intimidating.

Only connect . . .

> You must constantly be returning from the unfamiliar facts of science to the familiar facts of everyday experience . . . I think that popular science can be of real value by emphasizing the unity of human knowledge and endeavour, at their best. This fact is hardly stressed at all in the ordinary teaching of science, and good popular science should correct this fault, both by showing how science is created by technology and creates it, and by showing the relation between scientific and other forms of thought.
>
> J.B.S. Haldane, biologist

Many writers have used this technique to good effect. Darwin's *Origin of Species* launches its revolution with a discussion of domestic animals. Tom Waters, writing for the magazine *Discover*, cushioned the impact of the fifth and sixth forces by discussing how they affect what we already know, and he throws in a few familiar names to make us feel at home.

> What would it take to convince you that Newton's law of gravitation is wrong? That's a question many physicists have been asking one another lately. The law that explains why planets stay in their orbits and why apples fall from trees has been coming under attack. Some researchers say that Newtonian gravity, as updated by Einstein, has to be modified again; others claim that new and distinct 'fifth' and 'sixth' forces have to be added to gravity and the other three we know about already (electromagnetism and the strong and weak nuclear forces). Whichever way you phrase it, the important point is this: recent measurements of gravity have convinced some physicists that something is going on that neither Newton nor Einstein ever dreamed of . . . [1]

If what you have to say is so far outside of common experience that you can't find a link, remember that science is done by people, and that people can be very interesting. Science, like soap opera, is awash with controversy, danger, competition, scandal, adventure, personalities and big prizes. A mystery thriller with a little science in it is a lot more absorbing than a lot of science with no mystery or thrills.

Word pictures

At the start of her novel *Persuasion*, Jane Austen explains that Sir Walter Elliot only ever picked up one book, which was the list of Baronets. It always fell open at the same page:

> ELLIOT OF KELLYNCH HALL. Walter Elliot, born March 1, 1760, married, July 15, 1784, Elizabeth, daughter of James Stevenson Esq. of South Park in the county of Gloucester; by which lady (who died 1800) he has issue, Elizabeth, born June 1, 1785; Anne, born August 9, 1787; a still-born son, November 5, 1789; Mary, born November 20, 1791.

That passage provides a 'description' of Sir Walter, but it certainly doesn't provide us with a picture of him. It is an example of how, at one extreme, description can amount to nothing more than a dull recitation of facts which have meaning for the writer – as a shopping list does for the shopper – but holds no interest for anyone else. At the other extreme, description can be a thrilling invitation to step into a secret world.

Description, says Webster's dictionary, is 'discourse intended to give a mental image of something experienced (as a scene, person or sensation).' The creation of an image in the mind of the reader is by far the most effective way of engaging attention. Think back about books you've read and enjoyed – your memories of them will consist almost entirely of pictures, and yet the books contained only words. The more abstract your subject is, the more difficult it is to find visual images (ever seen a philosophy book with pictures?) but the more important it is to try. Some words have purely visual associations – colours for example – while others, such as 'summer' or 'twilight', bring sensations too. Look at this passage:

> The first frost of autumn was melting over the grey lawn when we arrived at our new laboratory. That was back in 1973. We ran the first tests before we unpacked our bags.

This provides a mental picture – though every reader's picture will be slightly different – and a setting for the science which follows. It compares rather well with the academic version, which might begin: 'Tests conducted in 1973 resulted in . . .', which conjures up nothing whatsoever.

Analogies are a powerful descriptive tool, and are as useful in writing about science for the public as they are in communicating with other scientists. When we think of the heart as a pump and of the atom as a solar system, we are making analogies. The biologists have their 'gene pool' and 'food chain', and physics would be floundering without its waves and fields.

There are different types of analogy. Similes are comparisons which usually include 'like' or 'as'. 'He drinks like a fish' and 'he is as sober as a judge' are popular similes, though both are open to scientific doubt. A metaphor, on the other hand, dispenses with comparisons and turns one object or action into another. A simile says that the technician is as tough, bright, enduring and valuable as a diamond; a metaphor says that the technician is a diamond, and leaves the reader to interpret the comparison.

Metaphors are more difficult to handle than similes. Many common ones have lost all meaning, even to some of the people who use them. 'To ride roughshod over', 'ring the changes', 'play into the hands of' and 'axe to grind' bring no picture to the reader's mind. Even if the figurative meaning of these phrases hadn't taken over entirely from the literal meaning, images of non-slip horseshoes and axe-grinders mean little to modern readers. Analogies work best if they are able to relate science to everyday contemporary events, which need in turn to be familiar to readers. We are all told at school that hydrogen sulphide smells like rotten eggs, but most people nowadays have never smelt a rotten egg.

Even when they are fresh and apposite, descriptive phrases can still provoke the wrong reaction, particularly if the literal and figurative get mixed up ('She sat with her head in her hands and her eyes on the floor'). A metaphor should call up a visual image: if none is produced, or if images clash, then the reader will perceive that you haven't seen the image yourself. You can't be crushed by a torrent or drowned by a landslide. A similar problem occurs if the imagery gets out of hand. 'A red herring has been dragged across the landscape of statistical mechanics' conjures up surreal visions which have nothing whatsoever to do with statistical mechanics, while 'when the race for the peaks began, Professor Fox was in the van' gives the immediate impression to those of us for whom 'van' is a vehicle that Professor Fox wasn't in the race at all. This sentence also sounds like it escaped from an epic poem: unintentional rhymes can be distracting, but are easily detected by reading your work out loud.

Whatever style you choose, there is always a place for description. This paragraph opens a hard-nosed adventure story about the darker side of botany: orchid smuggling.

> Rothschild's slipper orchid, *Paphiopedilum rothschildianum*, is arguably the most spectacular flower on earth. From a pyramid of mottled leaves rises a pair of thin, black stems. Nothing about the surrounding foliage gives any hint of the glory of the flowers which will suddenly burst out of this background. The eye falls first on the raised sepal head-dress of russet stripes. Only then do you notice the central protruding lip of bright veined purple or its flanking petals drooping in the manner of a moustache framing the face of an oriental sage. Like a piece of Ming porcelain, the flower has an unworldly delicacy, yet there is something unsettling, almost sinister, about its exotic perfection.[2]

This passage puts to good use a number of the literary devices which constitute good description. It is full of colours: black, russet, purple. There are metaphors: the leaves are a pyramid, the sepals a head-dress and the central petal a lip. We have similes: the petals are like a moustache and the flower like Ming porcelain. We learn about the flower's shape through words like surrounding, raised, flanking, framing, protruding, drooping. We are given a hint of the story which follows: the flower is sinister and unsettling. Yet the botanical description reads like this:

> Inflorescence: terminal, one of several flowers; flowers spiral or distichous. Flowers: small to rather large, resupinate, with an articulation between ovary and perianth in the genera with conduplate leaves; lateral sepals united; lip deeply saccate; lateral anthers fertile, median anther sterile . . .

Providing you are not a botanist about to set out on an expedition to find this orchid, which description would you rather read?

Historical reconstruction using descriptive techniques can be effective, particularly if you want to examine the development of a piece of research or if the topic is itself a historical science such as geology, archaeology or paleontology. Books such as Hugh Miller's *Old Red Sandstone* and Jacquetta Hawkes's *A Land* are full of vivid pictures in which imagination, combined with factual evidence, can bring us nearer to the truth about lost worlds.

Natural history, with its wealth of visual images, provides ample opportunity for description. Nature is forever on the move, and as Darwin shows in this passage from his journal, the picture needn't be static.

> With a stem so narrow that it might be clasped with the two hands [the Cabbage Palm] waves its elegant head at the height of forty or fifty feet above the ground . . . If the eye was turned from the world of foliage above, to the ground beneath, it was attracted by the extreme elegance of the leaves of the ferns and mimosae . . . In walking across these thick beds of mimosae, a broad track was marked by the change of shade, produced by the dropping of their sensitive petioles.[3]

By sprinkling verbs of movement (clasped, waves, turned, dropping) in the passage, Darwin has transported us through the forest: the effect is almost cinematic.

As far as description is concerned, the biologists have it easy. What does the theoretical physicist do for pictures? George Gamow solved the problem in his Mr Tompkins stories: Mr Tompkins lives in a world where quantum and relativistic events occur quite visibly in everyday orders of magnitude. But on a less ambitious scale, even if we cannot describe physical processes in terms which encourage visual images in the reader's mind, we can at least describe the events which record these invisible phenomena. Take as an example X-ray crystallography. Even if you could look inside the machinery, there wouldn't be anything to 'see'. So what do you describe? Actually, there is plenty there to interest the reader. What is the lab like? How easy is the apparatus to handle? How does the researcher feel about her work?

> I was shut up in a tiny room for days on end, pounding white crystals to dust. The pestle and mortar seemed out of place in the high-tech laboratory, and unbearably clumsy beside the fine glass tubes – tubes that would crumble between your fingers – which held the powder in the machine. Inside the machine was a revolving stage, and once I'd pressed and persuaded the powder to creep into the tiny tube it had to be mounted exactly at the centre of the stage, in line with the X-ray beam. Then the lights go out; the film is mounted; the high voltage supply takes over and the X-rays shine green on the tiny fluorescent disc on the front of the machine. A faint line across the screen is good news: the tube is in the beam. Today I would get a result . . .

Get the picture? When writing about science it is important to remember that it is important to record not only what is happening, but also what you see and feel.

Good for a laugh?

Isaac Newton, it is said, laughed only once in his life, when he was asked the use of Euclid's geometry. Ever since, scientists have been thought of as a rather gloomy breed who cope with the trials of life by disappearing into eccentricity. This can, however, make them comic: miserable Newton made his colleagues laugh because he was so absent-minded. He once walked on regardless for five miles after the horse he thought he was leading had slipped its bridle.

What the . . . ?

A singular man had visited Mr Frank Harris in *The Saturday Review* office and had asked to have the reviewing of scientific works for that periodical confided to him. Mr Harris said, Hell, Damn, Blast, Bloody, why don't you write some funny stories about Science? I don't know which of the elaborate expletives he used, but he used some.

Ford Madox Ford, writer and editor

While posterity has not recorded many jokes about the sciences, scientists commonly deal with phenomena that are extraordinary and people who are unbelievable – in short, they have to hand all the ingredients for a comic story. Yet scientists seldom convey these all-important aspects to those beyond the laboratory, perhaps under the misapprehension that if people start laughing at science, they'll stop trusting it. This is poor logic, for while almost everyone knows a doctor joke, people have more faith in doctors than in any other profession. No one will think less of you as a scientist for being funny, and you will do the profession plenty of good by knocking some of the stuffing out of it. Every time someone laughs at science, it becomes a little less 'Newtonian'.

Form and function

Norman Mailer said that form is a substitute for inspiration. This sounds rather damning, but anyone who's ever suffered from a shortage of inspiration will know how useful a good substitute can be. Form is rather like the scientific method: it gets you by until inspiration strikes. After plodding for months or years through reams of data the result you've expected for so long takes you by surprise, and seems to come almost by accident. When this happens, you know it. Robert Oppenheimer said that at such times the hair stood up on the back of his neck. Robert Graves said the same thing about chancing across a beautifully crafted poem, and the analogy is a good one: by working at a craft, be it the use of form in writing or the use of method in science, you can achieve something that transcends the prescription.

You might think that there are as many ways to write as there are subjects to write about and authors to write. This is true, but on the other hand, when it comes to actually writing something yourself you may find that not a single one of the many possible approaches suggests itself – in which case, you might find one of the following useful.

Tell it as it happened

A chronological approach is useful if you have a lot of information, several different locations and a number of different characters to introduce – the continuous thread of time serves as a guide to readers who might otherwise get lost in the action. James D. Watson's bestseller *The Double Helix* is an excellent example of straightforward storytelling: part history, part thriller and part adventure, it recounts a complicated series of events in America, Cambridge and London and yet maintains the continuity of the story throughout.

The basic 1-2-3-4- story might turn out to be rather ordinary, so experiment a little. Perhaps you could start with the end, and then tell the reader how you got there. This next passage, however, gives a simple step-by-step account of the first operation under ether in Britain, but keeps up the suspense by telling the story not as it happened, but as it is happening. Written in the present tense, it reads like an eyewitness account.

> A firm step is heard, and Robert Liston enters – that magnificent figure of a man, six foot two inches in height, with a most commanding expression of countenance. He nods quietly to Squire and, turning round to the packed crowd of onlookers, students, colleagues, old students and many of the neighbouring practitioners, says somewhat dryly, 'We are going to try a Yankee dodge to-day, gentlemen, for making men insensible.' . . . The patient is carried in on a stretcher and laid on the table. The tube is put into his mouth, William Squire holds it at the patient's nostrils. A couple of dressers stand by, to hold the patient if necessary, but he never moves and blows and gurgles away quite quietly. William Squire looks at Liston and says 'I think he will do, sir.' 'Take the artery Mr Cadge,' cries Liston. Ransome, the house surgeon, holds the limb. 'Now gentlemen, time me,' he says to the students. A score of watches are pulled out in reply. A huge left hand grasps the thigh. A thrust of the

long, straight knife, two or three rapid sawing movements, and the upper flap is made; under go his fingers, and the flap is held back; another thrust, and point of the knife comes out in the angle of the upper flap; two or three more lightning-like movements and the lower flap is cut, under goes the great thumb and holds it back also; the dresser, holding the saw by its end, yields it to the surgeon and takes the knife in return – half a dozen strokes, and Ransome places the limb in the sawdust. 'Twenty-eight seconds,' says William Squire. The femoral artery is taken upon a tenaculum and tied with two stout ligatures, and five or six more vessels with the bow forceps and single thread, a strip of wet lint put between the flaps, and the stump dressed. The patient, trying to raise himself, says, 'When are you going to begin? Take me back, I can't have it done.' He is shown the elevated stump, drops back and weeps a little; then the porters come in and he is taken back to bed. Five minutes have elapsed since he left it. As he goes out, Liston turns again to his audience, so excited that he almost stammers and hesitates, and exclaims, 'This Yankee dodge, gentlemen, beats mesmerism hollow.'[4]

The big adventure

Everyone recalls the childhood thrill of adventure stories in the company of the pioneers and pirates. For a while, as we snuggled down into a corner with a book, our hearts and minds were on the prairies and oceans with our heroes. Some of the most popular science writing has taken the form of adventures: Eve Curie's biography of her mother and Paul de Kruif's *Microbe Hunters* have long been popular favourites.

Feature articles are a good length for an adventure story. Lyall Watson's article 'Jungle Fever' gave readers a taste of an adventure which is usually overshadowed by the theory it produced.

Twice during the 1840s, Charles Darwin sat down and outlined his ideas about the formation of new species, then held back from publishing them, instructing his wife to do so after his death . . . In the Malay Archipelago, a different mind was at work on the same problem. Alfred Russel Wallace had already spent four years exploring the Amazon Basin, collecting and writing under difficult circumstances: lying ill with fever; watching his younger brother die from some unknown disease; crawling on all fours through the forests to the Brazilian city of Belém – only to be shipwrecked on his way home and to lose his samples and notes. Now, after eight years of travelling through the Indonesian islands, gathering more than a half-million specimens, Wallace was beginning to see sense in the tropical cauldrons he had been studying. While Darwin procrastinated, the younger explorer was immersed in the cut and thrust of evolution in action.[5]

In such stories it is tempting to skim over the science and dwell on the suspense. But even if you are writing about a fairly esoteric piece of scientific detective work, there is no reason to cut the science. Lots of research consists of thinking, and this process of problem solving can be compelling. Remember Sherlock Holmes: he used to spend hours, curled up in his well-upholstered armchair, puffing on a pipe, just thinking. Show the subject of your adventure story analysing problems and testing solutions, and include the dead-ends, mistakes, and moments of despair, for all of these add up to a three-dimensional picture of science in action.

Do-it-yourself

Most people aren't familiar with the paraphernalia of science, and yet they are surrounded by the phenomena that scientists would have us believe can only be observed on a computer screen or in a test-tube. If you can't bring the people to the science, take the science to the people. There is a great deal of physics and chemistry going on in a pan of potatoes, and a folded paper plane is a miniature text-book of engineering and aerodynamics. Do-it-yourself science comes in a number of guises: almost every activity from wine-making to dog-training offers the ingenious writer the opportunity to sneak in a little science. Sometimes there is no need to be surreptitious; a book on helping the children with school mathematics, working out how best to insulate a house or planning a garden would benefit from the authority of a straightforward scientific approach.

Personality piece

A newspaper editor used to impress on his journalists, 'There is nothing that can't be told in terms of people', and your article will remain unread or unappreciated if it does not acknowledge people as a well as science. Science is about objects and phenomena, but it is also about people – scientists and the public. What the public thinks about science and scientists has a lot to do with the way scientists present science: all too often it emerges as though it gave birth to itself. The human creator – you – is invisible in passive phrases such as 'it was discovered that . . .', or 'reagents were mixed . . .' They suggest, quite wrongly, that the creative self plays no role in scientific discovery. This effacement of self not only distorts the presentation of science, it also makes scientific writing difficult to read. Forget the passive and use active verbs: your story will flow more easily, and you'll find that interesting aspects of science, such as what people have actually done, will automatically reappear. Put the 'I' back into science!

First person

I almost always urge people to write in the first person. Writing is an act of ego and you might as well admit it. Use its energy to keep yourself going.

William Zinsser
Writer and professor of journalism

One features of the Bible which makes it so lively and accessible is its parables, stories teaching general truths through people. By attaching the science to people, just as novelists load ideas and situations onto fictional characters, you can introduce a plot, and, most importantly, give a sense of how research and technology affect people's lives.

Reader's Digest carried a long-running series describing the physiology and anatomy of various human organs – 'I am John's gall bladder', for example. Fortunately, bringing an aspect of science to life does not have to be done as literally as this. Since science is a human activity there are personalities involved in everything from the common cold to black holes. If you can't find a suitable candidate, invent one. ('Parr Wellbelow was having trouble with his handicap, trouble with his swing, and trouble finding a caddie strong enough to carry his clubs back to the bar. Then he discovered carbon fibre . . .'). If you can't involve Joe Public, have a look at the scientists involved, perhaps in the subject's past. The history of science is full of characters, and one of them is bound to fit the bill.

Tackling the issues

Often science is controversial or newsworthy or just plain interesting simply because it will change people's lives. Where scientific issues become social ones your job as a communicator is two-fold: you must account for both the science and its impact on society, and good science isn't always good social policy. Your readers may not be aware of the scientific aspects of a social problem, so you will have to convince them first of all that science has an important role in many controversial issues, and it is only by showing direct connections between the science and the problem that you will be able to do so. Articles like this aren't easy to write, and because so many factors are involved they often have to be rather long. Unless you feel passionately about the subject you will find that keeping up the interest is difficult. Start by defining and describing the problem, and then explore it in terms of people. Then, once your readers know why the subject is relevant to them, you can discuss the science.

Fighting talk

Almost every piece of science writing conceals an argument which is responsible for the set of ideas being explained. In any argument, however, you have not only to present your case, but also to explain the set of ideas which is being modified or overthrown. Sustained argumentative writing is difficult to achieve, and can be wearing for the reader. It is important therefore to be very careful about the structure of your argument, to carry the reader gently from one point to the next by using examples and illustrations, and to show respect for the opposing point of view. That said, demolishing objections is often more effective than substantiating your own position directly.

Beginnings and endings

Whilst the body of the text is important, the most important section of any writing, be it five hundred words in the local paper or a weighty textbook, is the opening paragraph. If the first few lines don't hook the reader you might as well not bother writing the rest. But once you've got over this first hurdle, and sweated and toiled to produce a stylish and enthralling middle, you have yet another difficult task ahead: finding a suitably final finish. No matter how brilliant your work is, an abrupt or lingering end can destroy the good impression you have made with the rest of the piece.

New beginnings

Some great writers have begun at a snail's pace. Marcel Proust introduced his great novel with endless ruminative pages on how he lay in his bed alone at night, thinking about this and that . . . But Proust was lucky: he published the result at his own expense and had no mundane commitments to interrupt his life of writing. You won't be so lucky, nor are you likely to present your thoughts to posterity, as Proust did, in ten stout volumes. A few thousand words, perhaps one slim volume, is the most that you can hope for, so get to the point as swiftly as you can. Writing a slow meandering start is the equivalent of putting a message into a bottle and hoping that it will wash up on an appropriate shore. Television and movies have borrowed the idea of a strong opening lead and often present at least part of the story before giving the title and credits. This doesn't mean that you have to offer a summary in the first line and then reiterate it throughout your text. If you rely too heavily on the old maxim, 'First I tell them what I'm going to tell them. Then I tell them. Then I tell them what I've said', you won't hold your readers' attention for very long.

Surprises, cliffhangers and mysteries are all good ways of attracting attention. You can open with a question – one which has an immediate relevance for the reader. Look again at the article on gravity we discussed on page 55. Questions naturally arouse our curiosity and start mental images flashing:

> Suppose you had discovered the perfect fuel oil substitute – efficient yet cheap. Would you sell the formula or market it yourself?

> What would you do if you were fired today? That happened to Dr Alan Seward 10 years ago, and he's never looked back.

Note that the accent is on personal and human openings: these always work better than objective-sounding ones.

You could try hitting the reader right between the eyes with a startling statement – have another look at the passage about the orchid on page 57. Or you could begin with a case history: highlight the real situation and use that to carry more abstract ideas.

> John Friar is seven years old, and like most boys of that age he likes football, ice-cream and television, and thinks his little sister is stupid. Yet John is unique. He is the first person whose badly broken leg was reconstructed using a revolutionary new alloy developed by the materials scientists at Highland Polytechnic . . .

You could lead with a hypothesis or your conclusion:

> We have long been told that people become what they think – that groups and societies are products of their individual and collective thoughts. It seems from recent research that people instead become what they *say*.

Here are some real examples of great starts. An account of recent work in particle physics shows just how powerful a well-constructed opening can be:

> Take a deep breath! You have just inhaled oxygen atoms that have already been breathed by every person who ever lived. At some time or other your body has contained atoms that were once part of Moses or Isaac Newton. The oxygen mixes with carbon atoms in your lungs and you exhale carbon dioxide molecules. Chemistry is at work. Plants will rearrange these atoms, converting carbon dioxide back to oxygen, and at some future date our descendants will breathe some in.
>
> If atoms could speak, what a tale they would tell . . .[6]

Here the authors of *The Particle Explosion* have managed to involve readers directly, given them cause to wonder, and then quickly proceeded to satisfying their curiosity.

> 'I was brought up to look at the atom as a nice, hard fellow, red or grey in colour according to taste', said Rutherford once.

So how did Rutherford come to change his mind? This surprising confession opens *The Infancy of Atomic Physics*, by Alex Keller.[7]

> I learned to pick locks from a guy named Leo Lavatelli.

Is this what we expect of a Nobel laureate? Perhaps, since it was Richard Feynman who wrote 'Safecracker Meets Safecracker'.[8]

> My father was on the top of a horse-bus, watching the festivities for Queen Victoria's jubilee on the night of June 22, 1887, when I was born in the house of my Aunt Mary . . .

This atmospheric and vivid picture provides a perfect setting for the early life of Julian Huxley.[9]

> Thirty miles south of San Francisco, an incredibly long, narrow structure knifes its way across the sprawling Stanford University campus. Motorists speeding along Interstate 280 pass right over this mysterious building, most of them oblivious to what is happening inside. The casual observer might well mistake it for a two-mile-long chain of boxcars rumbling through the scrub oak and manzanita of these parched California hills.

This paragraph has everything: a location, a mystery, and a moving picture complete with scenery and weather. Yet it introduces Michael Riordan's 'true story of modern physics', which reveals the usually esoteric world of high energy physics.[10]

In at the finish

Bringing your work to a graceful conclusion can be as difficult as getting started. The soundest advice on this matter comes from the King of Hearts in *Alice in Wonderland*: 'Begin at the beginning, and go on till you come to the end; then stop.'

Last words are important: they leave a lasting impression. J.D. Bernal leaves his readers thinking – he finished with a question:

> Having seen [the future], are we to turn away from something that offends the very nature of our earliest desires, or is the recognition of our new powers sufficient to change those desires into the service of the future which they will have to bring about?[11]

Robert Williams and Philip Cantelon conclude *The American Atom* by issuing a challenge:

> The American atom is history. It is also the greatest challenge to our future survival on this planet. We must master its past if we are to control its future.[12]

The last sentence of *The Origin Of Species* places Darwin's theory in a cosmic context. The final word could not have been better chosen:

> There is grandeur in this view of life, with its several powers, having been originally breathed into a few forms or into one; and that, whilst this planet has gone cycling on according to the fixed law of gravity, from so simple a beginning endless forms most beautiful and most wonderful have been, and are being, evolved.[13]

Rewriting

Even if you've chosen your format, selected your style, achieved an order, started intriguingly, proceeded enthrallingly and finished with a flourish, you've still got plenty of vital work to do. When you think you have finished your article, put it away for a few days and then look at it again. You will find that you notice the faults more readily after a break, and your revisions will be more effective. If you don't cross out at least a third of your first draft you are probably not being ruthless enough. If you can't bear to change anything, get a friend to do it for you – a fresh eye will also help in spotting mistakes and ambiguities. If you get help, make sure your critic knows for which publication and for whom you are writing. Once you have made alterations, rewrite the whole piece from start to finish – you may have crossed out something in the middle, thereby making part of the introduction redundant. Rewriting from beginning to end is particularly important if you write on a word-processor and have been rearranging paragraphs. Work on a printed copy rather than on the screen: text always looks different on paper.

When you have rewritten once, put your work away for a couple of days and then rewrite it again. Think about your readers, and the publication you have chosen. Is the article the right length? Is it in an appropriate style? Have you made the subject interesting? Does the introduction grab the attention and lead the reader into the main body of the article? Is the ending neat and conclusive? Have you included some human interest? Are your quotations accurate, relevant and effective? Does the story flow easily, and are the transitions smooth? Is there too much in the passive, or too much technical language?

Next, find a friend who is representative of your chosen audience and ask for criticisms. Do your friend's questions show an understanding of the piece? Do they indicate that you have left out important information or avoided a controversial or topical issue? Finally, be tough on yourself. For every sentence and paragraph you've written, ask yourself: What am I trying to say? Have I made myself clear? Could I have put it more concisely?

> ***The acid test***
>
> We suggest that whenever anyone sits down to write he should imagine a crowd of his prospective readers (rather than a grammarian in cap and gown) looking over his shoulder. They will be asking such questions as: 'What does this sentence mean?' 'Why do you trouble to tell me that again?' 'Why have you chosen such a ridiculous metaphor?' 'Must I really read this long, limping sentence?' 'Haven't you got your ideas muddled here?' By anticipating and listing as many questions of this sort as possible, the writer will discover certain tests of intelligibility to which he may regularly submit his work before he sends it off to the printer.
>
> Robert Graves and Alan Hodge, novelists and critics

After all the time and energy you will have spent on your typescript, the last thing you will want is for it to go astray. Newsrooms are hectic: documents can fall off one person's desk and end up several feet away – a whole world away – before the day is up. It helps if you address the package to the science correspondent or editor by name.

Rejection doesn't mean that the article is bad; it may mean that you have misjudged the level of the article, or simply that there isn't room for it in that particular publication. F. Scott Fitzgerald collected 122 rejection slips before he was published, and not a few afterwards. Richard Bach's *Jonathan Livingston Seagull* was rejected eighteen times before Macmillan agreed to print a cautious 7500 copies, and within five years it had sold seven million copies in the US alone. So don't give up: if you have an interesting subject and can write about it in an appealing way, someone somewhere will see the merits of your work. If you decide to submit to another publication, amend the article where necessary – you may find that the editor who rejected it will offer some constructive criticism, especially if you ask for advice. Don't let editors hold on to your work unless they give you a firm commitment to publish. If they aren't going to use the article, they should let you know so that you can try elsewhere. Another publication may be pleased to have it.

Keep a copy of your work: you may be asked to make changes quickly or need to discuss the article over the telephone with your editor. Keeping a copy can also guard against the unforeseeable. It is said that Isaac Newton's dog Diamond once knocked over a candle, destroying the almost finished labours of many years. 'Oh Diamond! Diamond!', exclaimed Newton, 'thou little knowest the mischief done.' Be warned: the ghost of Newton's dog can appear in many guises, from an over-zealous cleaner to a computer malfunction. Protect yourself against mischief by keeping paper copies of everything until your work is safely in print.

Recommended reading

Theodore A. Cheney, 1983, *Getting the Words Right: How to Revise, Edit and Rewrite* (Cincinnati, OH: Writer's Digest Books). Practical and Comprehensive step-by-step guide.

Lester S. King, 1978, *Why Not Say it Clearly: A Guide to Scientific Writing* (Boston: Little, Brown & Co). Medically inclined, but useful for scientists of all descriptions. Chapter 3 on common grammatical errors and chapter 5 on editing and revising are particularly good.

William Zinsser, 1989, *Writing to Learn* (New York: Harper & Row). A clear and witty account of how to improve your writing, and why it is important to do so. Zinsser analyses examples of good science writing, and devotes chapters to physics and chemistry, natural history, geology and mathematics.

References

1 Tom Waters, 1989, 'Gravity under siege'. *Discover*, April.
2 William Dalrymple, 1989, 'Raiders of the lost orchids'. *The Independent Magazine*, 19 August.
3 Charles Darwin, 1983, *The Voyage of the 'Beagle'* (London: Everyman/Dent), p. 24.
4 F. W. Cock, 1915, 'The first major operation under ether in England'. *American Journal of Surgery*, **29**(suppl.), 98.
5 Lyall Watson, 1989, 'Jungle fever – a naturalist's revelation in the tropics'. *The Sciences*, May/June.
6 Frank Close, Michael Marten and Christine Sutton, 1987, *The Particle Explosion* (Oxford University Press).
7 Alex Keller, 1983, *The Infancy of Atomic Physics: Hercules in his Cradle* (Oxford: Clarendon Press).
8 Richard P. Feynman, 1989, 'Safecracker meets safecracker'. *'Surely You're Joking, Mr Feynman!'* (London: Unwin Hyman).
9 Julian Huxley, 1970, *Memories* (London: George Allen & Unwin).
10 Michael Riordan, 1987, *The Hunting of the Quark: a True Story of Modern Physics* (New York: Simon & Schuster/ Touchstone).
11 J. D. Bernal, 1970, *The World, the Flesh and the Devil* (London: Jonathan Cape).
12 Richard C. Williams and Philip L. Cantelon, 1984, *The American Atom: A Documentary History of Nuclear Policies from the Discovery of Fission to the Present, 1939–1984* (Philadelphia: University of Pennsylvania Press).
13 Charles Darwin, 1986, *The Origin of Species* (Harmondsworth: Penguin).

2.3 Words and Sentences

Jargon

Ask a journalist about his dealings with scientists and the chances are that his main complaint will be about jargon. This monster rears its head every time anyone talks about their work – some people even see the use of jargon as a mark of their professional status. Even simple words can mean quite different things to different groups and professions: 'proof', for example, means different things to a barman, lawyer, scientist, printer, engraver and photographer. Where slang is straightforward and emphatic – the American poet Carl Sandburg described it as 'language which takes off its coat, spits on its hands and goes to work' – jargon is the opposite. When slang says 'nope' or 'no way', jargon says 'the answer to the question is in the negative.'

Flapdoodle

> *Jargon*: gibberish, jabber, mere words, bombast, balderdash, palaver, verbiage, babble, inanity, rigmarole, twaddle, twattle, fudge, trash, nonsense, bosh, rubbish, rot, drivel, fiddle faddle, flapdoodle
>
> *Roget's Thesaurus*

It is not inappropriate that the words 'gibberish' and 'jabber' owe their origins to a scientist. Jabir ibn Hayyan, who was known as Geber, was an eighth century Arab chemist who is reputed to have written two thousand books. Gibberish, according to Samuel Johnson, is 'the mystic language of Geber, used by chymists.' Chemists aren't the only offenders, and since scientists don't often deal directly with the public, it is sometimes difficult to judge what is and isn't jargon. If you are not sure, look in a dictionary that contains less than 100 000 words. A dictionary of this size will have no room for jargon, so if the word you plan to use is not in the dictionary, use a word that is. To get an idea of how baffling jargon can be for the non-scientist, speak to a scientist from another field. If you are a botanist, ask a physicist to tell you about Brillouin zones. As you listen to tales of k-space and CP Hex, remember that this is how you sound when you talk to the public.

One of the reasons why many people find science difficult to understand is that scientists use familiar words in an unfamiliar way, and invent words of their own. Some scientific words have different meanings in different sciences (physical and biological 'nucleus', mathematical and physiological 'pathology') and different meanings or connotations outside of their scientific usage ('mass', 'atomic', 'cell'). As J.B.S. Haldane once noted, this is something that probably cannot be avoided:

> We start with some ordinary word, such as "hot", whose meaning we think we understand. On the breakfast table are a tablecloth, a knife, and a pot of mustard. The plain man says the knife is cold, the mustard hot, and the cloth neither hot nor cold. A physicist will say that none of them is hotter than the others.[1]

The physicist is right, but so too is Haldane's 'plain man'. After all, he certainly gets a feeling of coldness from the knife and a feeling of heat from the mustard. These everyday perceptions, like common sense, are not something the science communicator should ignore or dismiss as nonsense – they can instead be taken as a starting-point, perhaps even as a subject, for an article or talk.

If you must use technical language when you write for a general readership be sure to define your terms and to give commonplace examples. You may find that you can take advantage of the unfamiliarity of a term to discuss its etymology – dictionaries often provide interesting information on the derivation of scientific words. For example, before it described the space between two kinds of teeth, 'diastema' meant a musical interval, and 'gynaeceum', which now means the female organs of a flower, used to mean the women's portion of a house.

Very new words are best avoided: they may not become established in the language, and could have disappeared before your book is published. It is usually possible to find a short phrase of familiar English words which would do much better than such clumsy constructions as 'deregionalize' and 'non-fragmentary'. Try to use specific rather than abstract terms, so that your words can be visualized – write 'box' rather than 'container' and 'bus' rather than 'vehicle'.

A popular form of scientific jargon is the abbreviation. Sometimes it helps the reader to know what the abbreviation means: CCTV is closed circuit television. On other occasions, knowing may contribute little to understanding; knowing that PTFE means polytetrafluoroethylene is irrelevant if all you need to know is that it doesn't stick. However, readers will want to know what the abbreviation stands for, so if you use one you should define it the first time it appears. The academic habit is to write 'Bardeen – Cooper – Schreiffer theory (BCS) can be used . . .', but it can be more helpful to the general reader if you write the abbreviation into the sentence – 'Bardeen – Cooper – Schreiffer theory, which is usually called BCS for short, can be used . . .' – as people often overlook the contents of brackets.

Clearing the fog

If technical terms are introduced, then they must be explained or handled in such a way that the general reader gets to understand them. Jargon should not be used to fog over anything that is difficult to explain.

Richard Fifield, Executive editor, *New Scientist*

It is easier to write long and technical terms than it is to say them; on the page, Latin or Greek sounding words somehow look grander than Anglo-Saxon ones. But it is well to resist the temptation; 'plants' may have to be 'uprooted' but this is preferable to 'deracinating vegetation', and 'improve the snapdragon' is better than 'amelioriate the antirrhinum'.

Short and sweet

Sir Ernest Gowers, in his manual on plain words, wondered why people are so easily lured into polysyllables. Recalling H.G. Wells's engaging hero Mr Polly, who revelled in 'sesquippledan verboojuice', he suggested that we are inclined to be seduced by the air of grandeur, distinction and out-of-the-ordinariness that long words carry. The trouble with long words is illustrated by this extract on . . . well, who knows?

> Experiments are described which demonstrate that in normal individuals the lowest concentration in which sucrose can be detected by means of gustation differs from the lowest concentration in which sucrose (in the amount employed) has to be ingested in order to produce a demonstrable decrease in olfactory acuity and a noteworthy conversion of sensations interpreted as a desire for food into sensations interpreted as a satiety associated with ingestion of food.

The Lancet spotted this paragraph in an early draft of a Government training manual. Its authors saw the error of their ways, and their second attempt makes much more sense:

> Experiments are described which demonstrate that a normal person can taste sugar in water in quantities not strong enough to interfere with his sense of smell or take away his appetite.

Whatever the words are, most people prefer them to be short rather than long. A simple rule is not to use words you wouldn't say: 'terminate', 'utilize' and 'erroneous' mean 'end', 'use' and 'wrong'. These long words are no more than fancy dressing for simple statements; they can give a false air of scientific impartiality or certainty to material that may not merit either. Look at these two lists: if you find yourself using a word from the list on the left, think whether the word from the list on the right mightn't do just as well.

approximately	about
attempt	try
commence	start

concerning	about
construct	build
demonstrate	show
elucidate	explain
endeavour	try
fabricate	make
frequently	often
indicate	show
initiate	start
manually	by hand
multiple	many
optimal	best
perform	do
possess	have
provide	give
sufficient	enough
superior	better

If you find yourself using long words and elaborate phrases, remember that though a short phrase of short words may be longer than one long word, it will be easier to understand. Erasmus Darwin left the long words out of his medical encyclopaedia written in 1794 because 'a short periphrasis is easier to be understood, and less burdensome to the memory'. If you need to look up 'periphrasis' in the dictionary, you'll know exactly what he meant.

But even short words can present problems. Using words like 'nature', 'degree', and 'instance' can produce a sentence that is limp and unspecific. 'In this instance, the reaction was due in some degree to the nature of the reagents' tells us very little. Think about what you actually want to say before you search for the words with which to say it. You can use a dictionary when you write, but don't expect people to have one to hand when they read what you've written.

The dictionary can help you to chose between words which look similar but which mean different things. If your lion was 'restive' your readers may wonder why you fled – had it been restless they would have understood. What are 'alternative Thursdays'? Was the effect affective, or the affect effective? Is 'alright' all right? In 1981 David Shaw, the media writer of the *Los Angeles Times*, asked newspaper editors to list those words which were most commonly misused. The top twelve were:

dilemma
egregious
enormity
fortuitous
fulsome

hopefully
ironically
penultimate
portentous
presently
quintessential
unique

Should one of these words – or any word which you think might not be all it seems – crop up in your writing, a brief trip to the dictionary will be worthwhile. Manuals of usage are also useful: 'enormity', though it literally – and correctly – means 'largeness', is now only used to mean 'monstrous wickedness'. A sentence such as 'the enormity of the population led to recurring shortages of food' might lead readers to suspect the hand of a punitive god, rather than a simple case of demand exceeding supply. There are several manuals of usage available, and details are given on page 86.

Stringing words together, on the model of polymer chemistry, can be taken to confusing extremes: while 'square-wave frequency modulation' is decipherable if the hyphen is in the right place, 'resulting in serious head on accident potential' is baffling, and far less clear than 'possibly causing serious head-on collisions'. The reason why such strings of words are difficult for readers is that they contain nouns which behave like adjectives. 'Accident potential' is two nouns, but 'accident' is also the adjective which tells us what sort of potential we're talking about. Look at this example, which provoked a correspondence in *Science:* a student called a section of a thesis 'lizard ovary winter lipid level change'. That's six nouns – so what does it mean? Another letter to *Science* pointed out that a minor reorganization and a well-placed hyphen could solve the problem: 'winter changes in lizard-ovary lipid levels' makes much better sense.

Superstrings

At the microscopic level . . . scientific articles are most commonly impaired by their authors' misuse of words, especially in the common practice of stringing nouns together to form multivalent adjectives. A 'B-cell', for example, is the name for a class of lymphocytes, a 'B-cell lymphoma' the name of a disease, 'B-cell lymphoma DNA' the name of DNA extracted from people suffering from this lymphoma, 'B-cell lymphoma DNA sequence' the nucleotide sequence of the aforementioned DNA, 'B-cell lymphoma DNA sequence non-homology' is a reference to the mismatch between the nucleotide sequence and another not yet specified . . . and so on.

John Maddox, Editor, *Nature*

Here's a sentence:

> In some of the hottest stars a series of lines known as the Pickering Series was discovered in 1896, which is spaced on precisely the same regular plan but the lines fall half way between the lines of the Balmer Series – not exactly half way because of the gradually diminishing intervals from right and left, but just where one would interpolate lines in order to double their number whilst keeping the spacing regular, though unlike the Balmer Series, the Pickering Series has never been produced in any laboratory: so what element was causing it?

It makes sense, if you look at it long enough – doesn't it? The problem with this sentence is that it is far too long. Actually when the physicist Arthur Eddington put these words together for a book he wrote in 1926, he organized them into four sentences: the longest had fifty-five words and the shortest six. Eddington, who managed to be a first-class scientist and a popular writer at the same time, knew that readers prefer their information in small parcels. Most of his sentences are sixteen, or twenty-two, or twenty-nine or eighteen words long, and only occasionally, when the idea really needs it, does he unleash anything longer. Eddington sometimes wrote very short sentences – six or seven words – which startle the reader to attention and make complicated ideas look very simple indeed. Here's his version:

> In some of the hottest stars a series of lines known as the Pickering Series was discovered in 1896. This is spaced on precisely the same regular plan, but the lines fall half way between the lines of the Balmer Series – not exactly half way because of the gradually diminishing intervals from right and left, but just where one would interpolate lines in order to double their number whilst keeping the spacing regular. Unlike the Balmer Series, the Pickering Series has never been produced in any laboratory. What element was causing it?[2]

Eddington's sentences were successful because although they varied in length, they were all short. Mostly we talk in short sentences, so a good test is to read your sentence aloud. If you have run out of breath or forgotten the beginning before you get to the end, the sentence is definitely too long.

Often you can shorten a sentence by replacing long phrases by shorter ones. Thus 'are disposed at places' becomes 'are placed', and 'has the capability of producing' becomes 'can produce'. Many phrases can be deleted altogether, such as 'it is worth pointing out that' or 'it is interesting to note that'.

Try to say everything once only. If you find yourself writing 'in other words' or 'another way of saying this is . . .', then go back to your original statement. Rewrite it more clearly, and you may be able to do without the extra explanation. Other repetitions are less easy to spot. Look at this sentence:

The submerged parts of the structure had to be kept clear of the accumulated sediment which built up around them.

We know that sediment accumulates, and that 'accumulate' means roughly the same as 'build up'. The sentence actually says:

The submerged parts of the structure had to be kept clear of the accumulated accumulation that had accumulated around them.

'The submerged parts of the structure had to be kept free of sediment' conveys all the necessary information.

Churchill remarked of an opponent that he had a rare talent for compressing the maximum of words into the minimum of meaning. It is not a talent to emulate. Apart from space constraints on the published version, your readers will not want to plough through masses of words to get at the details. This means you must get rid of both redundant information and redundant words. For instance: 'In such an experiment one must realize that the velocity of the atoms is a continuously changing parameter' contains no more information than: 'In such an experiment the velocity of the atoms is continuously changing'.

Word chemistry

One way to deal with long sentences is to distill them, boiling off the unnecessary words and leaving the remainder intact.

Before distillation:	**After distillation:**
Such a process is **a** very rare **event.**	Such a process is very rare.
The fact of the matter is that no results have been obtained.	No results have been obtained.
The results were **of an** intriguing **nature.**	The results were intriguing.
The situation regarding the shortage of equipment is a serious problem.	The shortage of equipment is a serious problem. (better still: The shortage of equipment is serious.)

Thus distillation can help you to express the essence of a phrase more clearly. The process of condensation can also help reduce wordiness:

Before condensation:	**After condensation:**
The work was carried out on an experimental basis.	The work was done experimentally. (better still: We experimented.)
The building was fitted with laboratory facilities.	Laboratories were installed.
A survey of the area was conducted.	They surveyed the area.

A great many stock phrases which might crop up in your writing can easily be condensed:

a great deal of	becomes	much
at high speed	becomes	quickly
at some future point	becomes	later
at this point in time	becomes	now
due to the fact that	becomes	because

Stock phrases like those on the left are dangerous, because we use them automatically and without thinking about what we really mean. The novelist George Orwell recognized this problem, and wrote:

> Modern writing at worst does not consist in picking out words for the sake of their meaning and inventing images in order to make the meaning clearer. It consists in gumming together long strips of words that have already been set in order by someone else, and making the result presentable by sheer humbug. The attraction of this way of writing is that it is easy. It is easier – even quicker, once you have the habit – to say *'In my opinion it is not an unjustifiable assumption'* than to say *'I think'*.[3]

A teacher once asked her class to use a familiar word in a new way. One boy stood up and said: 'The boy went home with a cliché on his face'. When asked to explain, he said: ' "cliché" means "a worn-out expression".' And so it does. Cliché originally meant a printer's stereotype plate. If the printer required the same piece of text often, he set it up and kept it ready for instant insertion. Should you ever find yourself leaving no stone unturned as you explore every avenue, see if there mightn't be a more original way of putting it.

This bring us to the set-phrase, the combination of words that undermines the idea. 'True facts', 'serious emergency' – here emphatic words enfeebled by over-use are given a qualifier to revitalize them. Civil servants used to stamp priority items with 'Expedite'. After a while, this stamp came to be used so freely that they had to introduce a new stamp: 'Urgent'. That worked for a while, but the cycle was soon repeated; presently 'Urgent' items were lying neglected. 'Very Urgent' followed soon, and doubtless the time will come when documents will move from government office to government office stamped with 'Frantic'. In this case, a tired adjective replaced another tired adjective. Some words don't need qualifiers: facts are necessarily true and emergencies are always serious. Some words cannot – and therefore should not – be qualified. 'Somewhat ideal' and 'quite negligible' illustrate the point: can anything be only 'somewhat ideal' and still be ideal? Are there degrees of negligibility? There are many other words which cannot be qualified, including:

absolute	certain	complete	dead
devoid	entire	essential	extinct
excellent	fatal	final	full
fundamental	harmless	immaculate	immortal
impossible	incessant	infinite	invaluable
main	mortal	obvious	perfect
pure	simultaneous	ultimate	universal
unique			

Qualifying claims such as 'it seems probable that under certain circumstances one might be justified in suspecting' can be effectively replaced by 'we think that . . .'. Qualifying phrases like 'shall I say', 'somewhat', 'rather' and the 'not un-' construction only add to the fog, and suggest that you weren't really sure of what you were trying to say.

Sentence structures

Learning a language by eavesdropping on native speakers and struggling through their literature is a luxury few of us can afford. Mostly we rely on a crash course, a dictionary and a book of grammar. Lacking experience of a new language, we have to judge our efforts not by whether the sentence sounds right, but by whether it follows the rules. We look to see whether the tense is right, whether the verb agrees with the subject, and whether the adjectives agree with the nouns. What we are doing is thinking about our sentence not as a representation of an idea or picture, but as a collection of words, each with a specific role to play in the structure of the sentence.

We think about our own language in a quite different way – we actually can, and do, learn to speak and write by listening and reading. Surrounded from birth by the sounds and symbols of language, we don't learn it as a list of nouns and verbs. Like the person who can't describe the journey he makes every day, we think of our own language as a whole, rather than as a collection of details. In fact, we think about our language very little at all. We navigate on automatic pilot.

This is fine when things are going well, but is no help when problems occur. That is when a logical, well-organized map is required. You can read a sentence over and over again, and while you know that there is something wrong with it – it doesn't sound quite right – you often can't tell what the problem is unless you step back from the meaning for a moment and look at the sentence as you would if it were in a foreign language. Like the engineer who can fix a machine simply by replacing one tiny component, you can often repair a broken-down sentence by looking at it closely, pinpointing the fault and then putting it right.

Before you can do this, you have to know what each component of the sentence does, and this is where thinking about grammar can help. In principle, it should be easy to state the rules of grammar and emphasize their importance. Grammar, it has been said, is the backbone of speech and writing. It should therefore make perfect sense and be eminently logical, but instead it is a tricky, inconsistent thing – just as arbitrary as a skeleton, in fact. Why twelve pairs of ribs rather than thirteen or eleven? Why thirty-two teeth? The answer has to do with function and evolution, the same factors which explain the structure of language. While you can't solve every problem by referring to a textbook, there are some that respond well to grammatical attention, so some hints and guidelines about the functions different parts of a sentence perform may be useful.

Verbs, subjects and objects

A verb is what makes a sentence. 'Help!' is a sentence. 'Help me!' is a sentence – it has a verb and an object. 'I'll help you' is a sentence – it has a subject, a verb and an object. It is the way in which you link subjects and objects by verbs which makes a sentence a success.

Verbs provide the action and carry the story. Here are the verbs from a paragraph of a story: travelled, will be, went, wrapped, slept, were whipped, were shipwrecked, tramped, sold out. Now here are the subjects and objects: they, taverns, attribute, café singer, novelty, Mexico, clothes, shawl, beaches, Panama, islands, droppings, birds, jungles, snakes, beetles, harvesters, season, nothing, world, them. The subjects and objects are a senseless collage without the verbs, and the verbs on their own present actions without actors or stage. It is only when the two are brought together that they stop being mere vocabulary and become language. The passage reads:

> They travelled a great deal, seeking new taverns, for the highest attribute of the café singer will always be her novelty. They went to Mexico, their odd clothes wrapped up in the self-same shawl. They slept on beaches; they were whipped at Panama, and shipwrecked on some tiny Pacific islands plastered with the droppings of birds. They tramped through jungles, delicately picking their way among snakes and beetles. They sold

themselves out as harvesters in a hard season. Nothing in the world was very surprising to them.[4]

When you have a story to tell, the subjects and objects are provided for you. You can chose from among them, but you cannot change them, at least not if you plan to tell the truth. The mouse is not a lion, the bunsen burner is not a furnace, and your professor simply does not have a glass eye, a knowing smile and a boyfriend called Boris. Verbs, on the other hand, are one component of a sentence which allow you a little more flexibility, and give you an opportunity to interpret the story for the reader. In the passage they sought new taverns – they didn't go to them, or stumble across them. They tramped through the jungles, while they could have marched, rambled, roamed, roved, trekked, walked, plodded or trudged. Each of these verbs would have got them through the jungle, but each is different. Marchers proceed swiftly and in order; ramblers stray and stay a while, admiring the scenery; the roamer and the rover travel here and there, one aimlessly and the other on a great adventure; the trekker is at odds with nature while the walker simply puts one foot in front of the other; the plodders and the trudgers are both weary, but one is determined and the other reluctant. Yet our heroes tramped: it was hard work, but the jungle proved no impediment to them. Choosing the right verb isn't easy – we over-use the common ones and forget the rest. If you get stuck, use a thesaurus.

Most words come in a maximum of two varieties – singular and plural. Not so verbs, as anyone who has ever tried to learn a foreign language will know. This variety of tenses also exists in English, though anyone who has learned the language through a study of scientific papers might be excused for not knowing this. The tense you use is the reader's guide to the chronology of your story. 'Because the summer *had been* dry, we *found* few frogs' tells the reader that the dry summer happened before you went looking for frogs. 'Because the summer *was* dry we *found* few frogs' suggests that you went looking during the summer, which is different. Avoid the present participle if you can. 'The source *being* masked, we inserted the sample' could mean either 'because the source *is* masked, we inserted the sample', or 'after we *had masked* the source we inserted the sample', each of which tells a different version of the story. Try to use simple verbs – 'experienced an increase' and 'underwent motion' mean 'increased' and 'moved'.

Verbs should agree with their subject. Look at this sentence: 'The new collection of clocks were put on display yesterday.' 'Were' sounds right because it is next to 'clocks', a plural noun. But the sentence should read 'the new collection of clocks was put on display yesterday'; 'was' is correct because the subject of the sentence is 'collection', a singular noun.

Sometimes 'was' and 'were' are left out of a series of passive verbs. 'The sample was weighed and several fractions taken for examination', is awkward because 'fractions', as the subject of the second verb 'takes', needs 'were', not 'was'. It is better not to omit the second 'were' or 'was', as this can sometimes be confusing even when both subjects are plural. 'The motors were oiled and switched off' leaves the reader wondering whether the motors switched themselves off. The easy way to avoid problems with passive verbs is to use active ones: 'I oiled the motors and switched them off' is unambiguous.

Ambiguity in all its guises is a real quicksand. Because you know what you mean to say, ambiguities are very difficult to detect, and a critical friend can be very helpful. When you write 'the impurity level was impressive', do you mean it was impressively large or impressively small? Does 'increased by one half' mean that four becomes four and a half, or six? You can give completely the wrong impression quite unwittingly with ambiguous language – you may have been 'researching plant systems in the Amazon Basin' in Milton Keynes.

Be consistent, particularly in lists. Look at this sentence: 'Einstein enjoyed physics, classical music, and to ride his bicycle.' Since you wouldn't say 'Einstein enjoyed to ride his bicycle', the sentence would be much better if you wrote 'Einstein enjoyed physics, classical music, and riding his bicycle.'

Infinitives which are apparently unattached to a subject are sometimes tricky to spot, but they cause problems because the unstated, but understood, subject of the infinitive may not be the subject of the next clause. Thus, 'to perform the operation, the rhinoceros had to be sedated' is misleading because 'the rhinoceros' is not the subject of 'to perform'. Although it is unlikely that the rhinoceros could have carried out the operation itself – at least not while it was under sedation – you can guard against potential misunderstandings by using more active, conversational expressions which make the meaning clear. 'We had to sedate the rhinoceros so that we could operate' is much better.

Don't worry about splitting infinitives. If you feel the urge to colourfully explain or to sheepishly admit, do so – it is often more effective than the 'correct' form, which separates the verb from the object. 'She began to painstakingly catalogue the vast collection' sounds better than 'she began to catalogue painstakingly the vast collection', and much better than 'she began to catalogue the vast collection painstakingly'.

The pompous pedant

> When established idiom clashes with grammar, correctness is on the side of the idiom. Put another way, if sticking grimly to rules of grammar makes you sound like a pompous pedant, you are a pompous pedant.
>
> William Safire, columnist, *New York Times*

Problems with pronouns

Check that your pronouns and possessive adjectives agree with their noun. This is often a problem with nouns like 'the faculty' or 'the audience' which can be used as either singular or plural. 'The committee ordered the fish to be removed from the laboratory because it smelt dreadful' is ambiguous, but common sense tells us that 'it' refers to the fish. The next phrase might however lead to some confusion: 'The committee ordered the fish to be removed from the laboratory because it smelt dreadful. They were incinerated.' Once we've decided that 'it' refers to the fish, we can only assume that 'they' refers to the committee.

Possessive pronouns are tricky: it often sounds better to say 'whose' when what you really mean is 'of which'. 'The lizard in the work by Professor Roberts, whose diet consisted entirely of maggots' might raise a smile, while 'the diet of which' is clear but clumsy. The way out of the dilemma is to start afresh – 'Professor Roberts' lizard was fed entirely on maggots' is straightforward enough.

Another problem with pronouns is gender. In English, most of our nouns have no gender, which is easy on the pronouns – most things are simply 'it' or 'they'. The problems start when we are talking about people. What do you do when you don't know whether you are dealing with a 'he' or a 'she'? Some people get round this by referring to a person of unspecified gender as 'they', but others would raise their hands in horror, claim that 'they' is a plural pronoun and substitute 'he'. A consequence of this is that women disappear from, and feel excluded from, much of what is written about people, because if the pronouns are anything to go by, women are very rare indeed.

The English 'he' meaning 'he or she' was actually invented by a conspiracy of grammarians in the eighteenth century. Until that time 'they' was quite acceptable as a singular pronoun. The pioneer printer William Caxton wrote 'each of them should make themself ready', and Shakespeare wrote 'God send everyone their heart's desire'. The grammarians' recommendation that 'he' should be used when gender was unknown was even endorsed by an Act of Parliament in 1850: 'words importing the masculine gender shall be deemed and taken to include females'. While the practice of 'he' caught on in language, the practice of 'including females' wasn't always so popular: in 1879 the Massachusetts Medical Society refused to admit a woman doctor because their rule-book referred only to 'he'.

'He' deceives

> Language is our means of classifying and ordering the world . . . and if it is inherently inaccurate, then we are misled. If the rules which underlie our language system, our symbolic order, are invalid, then we are daily deceived.
>
> Dale Spender, critic and historian

Often we can't know who people are, and so can be excused for not knowing whether they are male or female. However, this does not mean that we can assume that they are all male. If you put up a notice by the side of your archaeological dig which reads 'Anyone who wants to visit the site should report to the storeroom, where he will be issued with a hard hat', women might be entitled to wonder whether they were allowed on the site at all. ' . . .where they will be issued with a hard hat' is better. Sometimes 'he or she' (not 'he/she') serves as well as 'they', though it doesn't read well if you have to use it often.

Whatever the parliament of 1850 told us, there are many words in English which look exclusively male. Attempts to reset the balance have not always been successful, but many of the 'new' words are older than you might think. The word 'chairwoman' sounds clumsy to some, and many believe it to be a recent invention. Actually it was first recorded in 1685, only thirty years after the first recorded use of 'chairman'. Since then most chairs have been occupied by men, so the chairwoman died of neglect – just like the washermen, whose trade was taken over by women.

Many of the words which look as though they are about men are actually about people. There are often less exclusive alternatives:

men	people
workman	worker, labourer
manpower	labour force
mankind	humanity, the human race, people
manned	staffed, operated

If you do differentiate between the sexes when you write, do so equally. Nurses are male and female, so if you say 'male nurse' the reader is entitled to assume that you are making an observation which refers to his maleness, rather than to his being a nurse. If you must write 'male nurse', write 'female nurse' too. It is usually better just to write 'nurse'. Don't write: 'The control group consisted of 47 people, of whom 35 were women' – write 'The control group consisted of 12 men and 35 women', or even '35 women and 12 men'. Do not write 'mother' if you really mean 'parent', nor 'forefather' when you could use 'ancestor'.

Whatever you feel about equality, the fact is that many people feel irritated or excluded by the gender bias in our language. As a communicator, your job is to involve the reader in your writing, and before you can do this you have to be able to spot your own biases. If you can't, get a friend to help. Some publishers have produced guidelines on how to avoid gender bias: these are available from the addresses on pages 173-4.

Conjunctions

Conjunctions can be extremely useful for giving structure and continuity to your writing, but should be handled with care. Some define the logical relationship between phrases ('conversely', 'similarly', 'thus', 'therefore'), while others define a time sequence ('next', 'secondly', 'afterwards', 'subsequently'). They can provide the framework for an argument ('however', 'on the other hand', 'but', 'nevertheless'). It is important to remember that conjunctions must relate the phrase which precedes them to that which follows. Make sure that 'secondly' doesn't refer to your third piece of evidence, and that you have made one distinct statement before you discuss what's 'on the other hand'.

Punctuation

In writing, as well as speaking, you have a silent partner – punctuation. Like conjunctions, punctuation gives writing a structure and an internal logic. There are formal rules for the use of punctuation, but the best guideline to follow in practice is whether the punctuation you have used makes your sentence clearer and easier to read. Try thinking of punctuation marks as pauses of varying length, the longest being the full stop, then the colon, then the semicolon, and lastly the comma. Read your sentence aloud with pauses instead of punctuation, and see if it makes sense.

The right direction

> In its first capacity, punctuation is a traffic policeman averting chaos in the flow of prose; in its second, a good stage director supplementing the playwright's text with the weight of implication.
>
> John Simon, critic and columnist

The colon is used to indicate that what follows it is an example or explanation of what precedes it, as in 'Carbon is a group IV element: its electronic configuration is $1s^2 2s^2 2p^2$. However, you might prefer to write 'Carbon has the electronic configuration $1s^2 2s^2 2p^2$, and so is placed in group IV'. Indeed, unless you are using the colon to introduce a very long list ('Transition metals abound: scandium, titanium, vanadium . . .'), you can do without colons altogether. If you find you use them often, check to see whether the phrases they separate mightn't be better as separate sentences. There is never any good reason for more than one colon per sentence.

The semicolon is usually used instead of a conjunction, or instead of one conjunction where you would otherwise use two. For example: 'Diamond has a covalent tetrahedral structure *and* is the hardest substance known' could be written 'Diamond has a covalent tetrahedral structure; it is the hardest substance known'. Similarly, 'This plant relies on high levels of sunlight, *and consequently* could not survive an extended winter' becomes 'This plant relies on high levels of sunlight; consequently it could not survive an extended winter.' Semicolons are useful for separating lists, as in 'Transition metals fall into groups of three: Sc, Y, La; Ti, Zr, Hf; . . .' Like colons, semicolons are rarely indispensible, but they can be used occasionally to good effect.

Commas divide a sentence into a series of phrases. Place the commas so that the sentence makes sense and then read it aloud, pausing at each comma. If that doesn't sound right, read the sentence without commas and see where you pause. Watch out for ambiguities: 'The crabs which weighed over 500g were taken home for supper' implies that there were crabs weighing less than 500g which got left on the beach. 'The crabs, which weighed over 500g, were . . .' (two commas) means that all the crabs weighed over 500g. If you stop and think about commas each time you come across 'which', you should be able to eliminate ambiguity from this source.

Brackets (parentheses) are useful for giving supplementary or explanatory information, or for separating a brief diversion from the rest of the sentence: 'The lichens (pronounced 'lykens') are an excellent example of symbiosis'. Try not to use brackets too often – they make sentences look more complicated than they really are. Try a pair of commas instead, or a pair of dashes – thus – which emphasize the word or phrase they enclose. If your brackets contain only one word or one complete sentence – ('The (blue) crystal is hydrated.') – decide whether or not the word or phrase they enclose is necessary, and, if it is, whether it really needs the brackets. Always make sure that the sentence makes sense without the phrase in brackets – you should be able to remove the contents of the brackets and still have a complete sentence. If you use brackets to provide an explanation, make sure you put it in the right place. 'In 1988, private hospitals were supplied with more than 40 000 units (about ⅔ of a pint) of blood' illustrates the problem. In this case it would be better to abandon '40 000 units', and say '27 000 pints'.

Dashes should be typed as a hyphen with a space on either side. The most valuable use for the dash is to indicate the breaks and pauses and sudden changes of direction which occur so often in everyday conversation. 'Professor Jenkins insisted – and when he insisted no one argued – that we arrive for work at 8 a.m. sharp.' As with brackets, make sure that the sentence is complete without the words between the dashes. You can also use dashes to separate a list from a sentence about the list: 'Einstein had a number of interests outside physics – politics, music, philosophy and cycling.'

Question marks are relatively trouble-free, although some people find rhetorical questions irritating. Exclamation marks are a different matter; adding one to a sentence doesn't automatically make that sentence surprising or emphatic, and if you have written a surprising or emphatic sentence you won't need the exclamation mark. They are useful in dialogue or quotations because they indicate a tone of voice, but otherwise they are best ignored.

Don't worry about small matters such as the number of quote marks to put round quotes or whether to put points after initials, as publications have their own style for such things and will expect to have to change yours themselves. Your job as a writer is to get the sentence structure right, and you can do that with (, ; : - .). Provided that your meaning is clear, the copy-editor will fix errors of punctuation – that is, if 'punctuation errors' even exist. Punctuation is not so much right or wrong, but bad or better. Forget the rules and aim for clarity: the punctuation is good if the sentence makes sense.

Recommended reading

American Psychologist, 1975, 'Guidelines for nonsexist use of language', **30**, 682-6.

Sir Ernest Gowers, 1987, *The Complete Plain Words* (Harmondsworth: Penguin). A poorly organized and frustratingly incomplete book, but once you find what you want, Gowers is a clear and reliable guide to saying what you want to say as simply and effectively as possible – for UK English.

Reader's Digest, 1985, *The Right Word at the Right Time: A Guide to the English Language and How to Use it* (London: Reader's Digest). The best sourcebook available. A comprehensive and thorough guide to usage and style, including the differences between British and American English.

Matt Young, 1989, *The Technical Writer's Handbook: Writing with Clarity and Style* (Mill Valley, California: University Science Books; distributed outside North America by Oxford University Press). Who or whom? That or which? A while or awhile? This book, written by a scientist and journal editor, answers these and many other questions. Entries cover such topics as jargon, metric units, figures and tables, as well as grammar, style and organization.

The Oxford Dictionary for Writers and Editors. Arbitrates in spelling disputes; very useful for abbreviations, proper names, and new or colloquial words.

The New Collins Thesaurus. This is organized like a dictionary, and so is easier to use than the better known *Roget's Thesaurus*.

References

1 J.B.S. Haldane, 1985, 'What "hot" means'. *On Being the Right Size,* edited by John Maynard Smith (Oxford: Oxford University Press).
2 A.S. Eddington, 1942, *Stars and Atoms* (London: Oxford University Press), p. 61.
3 George Orwell, 1968, *Collected Essays, Journalism and Letters,* Vol. IV, edited by Sonia Orwell and Ian Angus (London: Secker & Warburg), p. 134.
4 Thornton Wilder, 1982, *The Bridge of San Luis Rey* (London: Penguin), p. 84.

2.4 Numbers

Mathematical language doesn't mean much to non-scientists, often because scientists use numbers which mean very little. An engineer interviewed in the wake of the Challenger accident said that safety measures had since been increased by an 'order of magnitude: 10 to 100 times'. Common sense says there is a big difference between 10 times and 100 times, and that 'safety measures' aren't a quantity to be multiplied. By adopting 'scientific' terminology, the engineer sounded scientific, but didn't make sense.

Smelling a rat

> 33⅓% of the mice used in this experiment were cured by the test drug; 33⅓% of the test population were unaffected by the drug and remained in a moribund condition; the third mouse got away.
>
> Related by Erwin Neter,
> Editor-in-chief, *Infection and Immunity*

Quantities

Most people find very large and very small numbers difficult to grasp. Saying that the universe is huge or that protons are tiny doesn't tell us much, but comparing one to the other, or better still, each to some familiar object, throws them into a context which is easier to imagine. Comparisons are often useful when you have to deal with quantities that are not on an everyday scale. Look at this statement: 'The Moon is 240 000 miles away from the Earth'. How far is that? It's two and a half times the distance from Greenland to South Africa. It's the distance you'd cover if you walked for five years. How about this one: 'Atoms are small – so small that when chemists count out quantities they work in 'moles', each of which contains six hundred thousand billion billion atoms.' If chemists need six hundred thousand billion billion atoms before they can handle them they must be pretty small . . . but how many is six hundred thousand billion billion? It's actually about the number of apples you'd need to fill a bag as big as the Earth.

The quality of quantity

> Through and through the world is infested with quantity: to talk sense is to talk quantities. It is no use saying the nation is large – How large? It is no use saying that radium is scarce – How scarce? You cannot evade quantity.
>
> Alfred North Whitehead, mathematician and philosopher

Comparisons can be used to give a particular interpretation to a piece of information: Jupiter's Red Spot can be described as '40 000 kilometres across' (how big?), 'ten could fit into the distance between the Earth and the Moon' (quite small), or 'three times the size of the Earth' (enormous!). This is an old trick: in the early days of microscopes researchers thought that they could see a perfectly formed miniature human being inside each spermatozoon. A popular medical writer of the time made it quite clear what he thought of this idea when he wrote that these tiny people 'must possess a much greater degree of minuteness than that which was ascribed to the devils that tempted St Anthony; of whom 20 000 were said to have been able to dance a saraband on the point of the finest needle, without incommoding each other'.

Numbers with lots of zeros after them are difficult to read. At a glance, the eye can only count up to four identical objects, so if you write more than four zeros your readers will have to stop to count them. It is better to write the number in words. Write 10 million, rather than 10^6. Even for people who understand the notation, 10^6 looks like a small number – smaller than 7985, for example. Numbers with lots of significant figures are difficult to take in, so even if the number is very definitely 14 328, write 'just over 14 000'. Use proper fractions rather than percentages – 'three quarters' rather than 75% – and if the number is small, try 'one in seven' rather than 14%.

The units you use can influence the way people perceive numbers. If your readers are young people they will expect you to use metric units, even for quantities such as travelling distances which most people still think of in miles. Older people will find quantities measured in grams, centimetres and Celsius difficult to imagine. When you present a quantity as a number and a unit, the size of the number will have more impact than the size of the unit. Thus 50 centimetres appears, at first glance, bigger than 5 metres. You can use this to your advantage – if you want a quantity to seem big use a large number with a small unit. Always define a unit by comparing it to a familiar unit, or to the size of a familiar object.

Words like 'exponential' or 'logarithmically' mean less to the non-scientist than 'steep' or 'very quickly', and what you lose in accuracy you gain in understanding. Avoid equations at all costs: they are the ultimate in scientific jargon, and the most abbreviated abbreviation imaginable. Chose a very small number of variables – two is enough – and describe their relationship in words. 'Its viscosity goes up as it cools' is much better than $\eta = D\exp(E/RT)$. Try to avoid using symbols for variables, even for common ones. Even though it is easier to write 'T' than 'temperature', the reader will still have to spend time translating T into temperature, and should not be required to make the extra effort.

Probabilities

Some people have great difficulty making sense of probabilities. Tell someone that there is a 50% chance of his first child being a girl and a 50% chance of his second child being a girl, and he may well conclude that the chance of one of his first two children being a daughter is 100%. Americans didn't travel abroad in the late 1980s, because in 1985 seventeen American travellers were killed by terrorists. The 17 who died were actually a tiny fraction of the 28 million Americans who went abroad in 1985 – that's one in 1.6 million. The chance of dying in a car crash is about 1 in 5300, yet the people who stayed at home wouldn't dream of giving up driving.

Our reactions to real situations in our personal lives is governed to some extent by a contradictory mixture of two attitudes: the first is 'it won't happen to me', and the second is 'oh yes it will'. However, one recent survey has found that around 80% of people asked clearly understood the implications of the statement 'a one in four chance' when asked about the chances of a hypothetical couple's children inheriting an illness.[1] Clearly people can make accurate assessments of probability; perhaps they are more likely to do so when the probability is happening to someone else. Nevertheless, horrors abound: a newspaper headline in 1973 claimed that vitamin C could 'reduce your chance of dying by 50%'.

The best way to communicate probabilities is to compare them to the probabilities of everyday events. The comparison of probabilities is especially useful in discussions of risk, particularly if you can compare the risk to one that people voluntarily or unwittingly run already. Often risks which cause wide public concern, such as the health risk from emissions from chemical and nuclear plants, are much lower than some which arouse less concern, such as the risk of being involved in a driving accident. Take as an example residents of a town twenty miles from a nuclear power station, who are worried about its effects on their health. You could explain the risk they run by offering the following comparisons:[2] the risk to their health is equivalent to

smoking 1.4 cigarettes
living for 2 days in New York
travelling 150 miles by car
eating 40 tablespoons of peanut butter
travelling 6 minutes by canoe
spending one hour in a coalmine
drinking 0.5 litres of wine

This account is far from complete, and your readers would be entitled to challenge you on a number of points. For how long do you have to live near the power station to run this risk? In this case, the calculation was for 150 years. What risk are we talking about? Here the calculation is based on increased chance of death. Each of the above increases your chance of dying in any one year by one part in one million. But, asks the reader, what is it that kills me? These figures give the increased chance of dying as a direct consequence of a particular activity, and a cause of death had to be chosen in each case. Spending one hour in a coalmine will increase your chance of death from black lung disease by one part in one million; spend three hours there and you run the equivalent risk of dying in an accident. A death caused by living in New York would be from air pollution rather than murder, and the peanut butter will give you liver cancer, not heart disease. But in life we cannot separate risks like this. How do the figures interact? How should a smoking, wine-drinking, car-driving canoeist who holidays in New York react to these figures? These are challenges which your readers are entitled to make.

Even if the readers accept your figures, they still have to grapple with the idea of 'increased chance of death in any one year by one part in a million'. A more manageable comparison might be one which compared loss of life expectancy. For example:[3]

Risk	Loss of life expectancy (days)
Being unmarried, males	3500
Heart disease	2100
Being unmarried, females	1600
Cancer	900
Road accident	207
Suicide	95
Murder	90
Drowning	41
Accident with a gun	11
Drinking coffee	6
Accident at a nuclear reactor (Union of Concerned Scientists)	2

(At this point the canoeist who goes to New York has given up worrying about the power station and has decided to get married.)

So reduced life expectancy is another way of comparing risks. But what do people make of it? Drowning reduces life expectancy by 41 days, but if you drown you die, and are not left around to contemplate a shorter life. If you drown tomorrow, your life will almost certainly be reduced by more than 41 days. Reduced life expectancy is a difficult concept – most people think that they will live long lives, so that the few days or even years that they might lose at the end of their lives seem a very long way off.

Although these methods present some difficulties, they do provide the public with a way of judging one risk against another, which is valuable. But the size of a risk is one thing: how people perceive it and whether it is acceptable is another matter. People accept risks which they run voluntarily (a skiing accident, for example) which are a thousand times greater than one imposed on them (pollution in drinking water). If you wanted to compare these two risks, you would have to take this difference of attitude into account.

While discussions of risk and probability are difficult, they are important if the public is to get a picture of what science and scientific knowledge are really like. Comparisons are very useful, but to be effective they must be with everyday risks and probabilities. The risk of 'Chernobyl' happening in Britain was described by a Cabinet Minister as about as likely as an aeroplane crashing into a football stadium during a cup final, which is a comparison few people are in a position to assess.

Recommended reading

J. Clarence Davies, Vincent T. Covello and Frederick W. Allen, 1987, *Risk Communication* (Washington DC: The Conservation Foundation). An animated and provocative account of risk communication, with contributions from scientists, policy advisers and journalists.

Darrell Huff, 1988, *How to Lie with Statistics* (London: Penguin). You know what you mean: this book will tell you what other people will think you mean.

John Allen Paulos, 1988, *Innumeracy: Mathematical Illiteracy and its Consequences* (New York: Hill and Wang). On the principle that understanding what people misunderstand is a first step to enlightening them, this book is an entertaining account of the misunderstanding of mathematics.

References

1 John R. Durant, Geoffrey A. Evans and Geoffrey P. Thomas, 1989, The public understanding of science. *Nature,* **340,** 6 July, 11.
2 R. Wilson, 1979, Analyzing the risks of daily life. *Technology Review,* **81**(4), 40.
3 B. Cohen and I.S. Lee, 1979, A catalog of risks. *Health Physics,* **36,** 707.

2.5 Illustrations

It may depress anyone with aspirations to authorship to learn that most people remember only 10% of what they read. However, they remember 30% of what they see, and 50% of what they read and see. Thus pictures are an invaluable aid if your aim is to leave a lasting impression.

Draw and describe

And ye who wish to represent by words the form of man and all the aspects of his membranification, get away from the idea. For the more minutely you describe, the more you will confuse the mind of the reader and the more you will prevent him from a knowledge of the thing described. And so it is necessary to draw and describe.

Leonardo da Vinci

Pictures

Illustrations are a useful device for clarifying ideas which are difficult to express in words. Try drawing an abstract shape on a piece of paper, and then describe the shape to a friend who has not seen what you've drawn so that she can draw the shape for herself. Chances are that however closely your friend follows your instructions, her drawing will look nothing like yours.

Illustrations are not simply ornaments, nor are they something to be added once the text is finished. They should be an integral part of the text, and prepared in parallel with it. Pictures can provide a valuable break for the reader, and can bring structure and variety to the text. But too many illustrations can be confusing. Apart from being swamped by the information, readers will have trouble following the text as it snakes its way between the pictures. If you want to use a number of pictures spread them through the text – if you mention seven pictures on the same page the reader is bound to have to turn over to look at some of them.

Pictures give emphasis to the text they illustrate. The reader's conclusions about the main theme of your article or the main points of your book should be confirmed by the pictures. If you illustrate minor points and not major ones, the reader may be confused.

The best place to use pictures is where words simply won't do, or won't do well enough. If you get stuck with a description or an instruction, think how you would communicate if your readers were in the room with you. If you can just about get your message across if it is accompanied by a vast amount of handwaving, a picture may well help. If you would stop talking and show your guests a model at a particular point, use a picture in your text. If you would demonstrate, use a diagram or schematic.

Pictures should be carefully chosen: they should be precise and specific. Don't overload the readers with information, and then expect them to ignore most of it. If you find yourself writing 'if you look at the small duct in the shadow of the vacuum chamber in the top left-hand corner of this photograph', your photo is not precise enough: find another one that only shows the duct, or trim your photo and enlarge it. In line drawings only include the essential features, and remove any detail which might distract or confuse the reader. Avoid cluttering a picture with too many labels. If there are several features which deserve a mention, give them numbers and discuss them in a caption.

Illustrations should be relevant, but should elaborate on rather than simply mirror the text. If you are writing about the friction between a tyre and the road, the reader will not be enlightened by the familiar sight of a parked car. Show the readers something they haven't seen before. Think whether what you are trying to show can actually be shown. If the phenomenon you are describing manifests itself as the motion of a needle across a dial, you will have trouble showing this in a still photograph. Show something: a photograph of a radiator does not illustrate the statement 'heat rises'. A line drawing would be better.

Don't leave illustrations to speak for themselves. If you have no reason to mention a picture in the text, it shouldn't be there. Even the most obvious picture deserves a mention.

Check with your publisher whether you can use colour photographs. Usually you will have to make do with black and white, so you will have to use some imagination if colour is an important feature of your picture. Black and white photographs should show good contrast, and you should give your publisher a print. Colour photographs should be provided as transparencies. Photographs should be originals, not ones that have been printed. A printed photograph is made up of a pattern of tiny dots with space in between them. This is called 'screened'. Screened photos don't look good when reproduced, and it is not possible to reduce or enlarge them.

Photographs have a quality which line drawings can never possess: they are accepted as evidence of the truth. A good photograph can take from you much of the burden of explanation. A good line drawing, on the other hand, can only illustrate: you still have to convince the reader that what you have drawn is an accurate representation of reality. While photographs present reality, they often contain irrelevant details which could distract the reader. Sometimes a line drawing is more effective.

Tell the reader which view you are presenting, and provide a familiar point of reference or comparison for questions of scale. See if you can get a tree into your picture of the radio telescope, and show a person at work in the accelerator tunnel. A superimposed line drawing can help out in difficult circumstances: draw a circle the size of the Earth in the corner of your photograph of Saturn. Otherwise, try verbal comparisons: 'ten of these laid end to end would fit across a human hair'. Magnifications and scale bars don't mean much to the uninitiated.

The background of your picture is very important. A nuclear power station photographed on a spring morning surrounded by daffodils presents quite a different image from one photographed on a bleak coastline in the rain. Even more so than words, pictures mean different things to different people – they see what they want to see. Your job is to make the readers see what you want them to see, so try out your illustrations on some friends and amend them where necessary.

If there are people in your pictures, remember that 50% of the world's population are women, and that a number of them are scientists, and that science is practised by and of interest to people of all nations and races. Illustrations should be accurate: someone is bound to catch you out if you use a photograph of an Indian elephant to illustrate a story about conservation in Africa. If your story is a contemporary one, make sure that your photograph is up to date. Even if the science in the picture is fine, it is worth checking the background. Does the billboard behind the bridge read 'I like Ike'? Is the technician wearing a pair of trousers that you would have thrown out with the beads and afghan coat?

There are some organizations which can offer advice and put you in touch with professional illustrators. Their addresses are given on page 173.

Graphs

Graphs should contain the minimum of information. How accurate do you need to be? Would a pie-chart or bar-graph be more effective? People perceive lengths more accurately than they do angles, so if your pie-chart would consist of a number of similar segments, use a bar-graph instead, where the relative values will be more apparent. On the other hand, pie-charts are more familiar to most people than graphs, and so look less intimidating.

The scale you chose for a graph is the equivalent of the background in a picture. Are you showing fluctuations, which would require a small spread of values on the vertical scale, or is your point that the parameter is steady on average, which would require a larger scale? Do you chose to include the origin? While you may save space by not including zero, you may give a false impression: in a graph the information is absorbed at a glance, and few people read the numbers on the axes. The horizontal scale will be limited by the width of the page. If the graph is for a magazine, you can chose whether it fills the width of one or several columns.

Printing and publishing

Once you've submitted your manuscript, your line drawings will probably be redrawn by a professional illustrator. The illustrator's job is to provide an accurate representation of your rough drawing, which means that he or she will follow the detail of what you have drawn exactly, and will probably be unqualified to supply details you have missed. If you draw a spider with only six legs because there wasn't enough room for the others, a big label which reads 'spider' will not be a big enough clue to the illustrator that two more legs are required. If your graph crosses the x axis at 0.1 because you missed zero, there it will remain. If the wires in your circuit touch where they shouldn't, they will touch in the illustration. If you feel that this degree of accuracy puts unnecessary strain on your artistic resources, you can write notes to the illustrator: 'this angle should be 90 degrees', 'these particles are all the same size' or 'this is supposed to be a rabbit' can all help. Put notes like these in a bubble, or they may be confused with labels. The illustrator probably won't read the text, so your note had better be specific – if you are writing about a partridge in a pear tree, you don't want a picture of an eagle in a giant redwood.

If you are preparing your own drawings, make them big so that they can be reduced photographically. This has the advantage for the amateur artist that any minor waverings will disappear, but be careful with lettering – if the letters are too small or your pen too broad the characters may run together or close up when they are reduced. Find out the size of the pages and test your lettering by making the appropriate reduction on a photocopier. In publishing, reductions are specified as fractions of lengths, not of areas. A picture reduced by 50% will be half as tall and half as wide, but one quarter of its original size.

Think about the page layout. Most pages are rectangular, and deeper than they are wide. This design is called 'portrait'. Check your publication to see how wide the columns of text are. If your photograph is much wider than it is deep, it is not going to fit onto a portrait page or into a single column of text unless it is very small. Wide pictures can be set across a number of columns in a magazine, or sideways ('landscape') in a book.

Some pictures however – those which need to be a particular way up, such as people or mountains – look peculiar sideways, and it is irritating for the reader to have to turn the page round to see them. Pictures in parts present a particular hazard for page designers – however hard they try, they will not be able to fit your three square photographs into one square space.

When you submit your manuscript, keep the words and pictures separate: words go to the typesetter and pictures to the artist or photographic department. Number the pictures and give each one a title, even if they won't have numbers or captions when they appear in print. This will help the editorial staff, who may otherwise may not be able to tell a beta particle track from a step defect. Mark on each picture which way is up. If your pictures aren't referred to explicitly or by number in the text, make it clear where each one should appear by making a note in the margin of your manuscript.

When you read your proofs, read the figures as carefully as you read the text. Check that all the necessary labels are in place, and that all the information has been provided which the reader will need to connect the illustration with the text. Make sure the picture doesn't appear before it is mentioned in the text, and don't let pictures interrupt the flow of the writing.

Pictures take up a lot of space and can be expensive. Whenever you decide to use a picture, think how many words it would take to convey the same information and to give the reader the same impression as you could with the picture. If you would fill less space with the words, don't bother with the picture. On the other hand, a carefully chosen picture can be excellent value.

Recommended reading

Elaine R.S. Hodge, 1989, 'Scientific illustration: a working relationship between the scientist and the artist'. *Bioscience,* **39**, February, 104–111.

Darrell Huff, 1988, *How to Lie with Statistics* (London: Penguin). Includes a section on how not to lie with graphs.

Edward R. Tufte, 1983, *The Visual Display of Quantitative Information* (Cheshire, CT: Graphics Press). An account of past and present practices in the representation of data, and a reminder of how easy it is to mislead.

2.6 Working with Publishers

Whichever form of writing you choose, your manuscript should be typed double spaced with wide margins on one side of A4 or quarto paper. If your work was unsolicited, or if you don't yet have the publisher's commitment to publish, send photocopies of your pictures and keep the originals until you are asked for them.

If you submit an article to a magazine or newspaper, the editor will let you know if he or she intends to publish your article, and will be able to tell you when it is likely to appear. You probably won't see your work again, or be paid for it, until it is published. The editor may telephone you if any queries arise when your work edited, but otherwise you won't have any contact with the publication. It is unlikely that you will be able to see your proofs.

If you have written a book, it's a different story. First, your manuscript will be sent to a small number of 'readers'. These will be people who, in the publisher's opinion, are qualified to judge whether you have written a book which is appropriate in terms of length, style and content for your intended readership, and which fulfils the requirements of your contract. The readers may recommend some changes, and if so the editor will discuss these with you. If all's well, your book will be passed for publication.

Editing

Once your manuscript has been passed for publication, it will be given to a copy-editor, who may be freelance. The copy-editor has two main tasks. The first is to make sure your typescript makes sense, both at the level of the individual sentences and as a whole. He or she will correct any spelling, grammar and punctuation mistakes, and may make stylistic changes. Apart from being a fresh eye which can spot mistakes you or your editor may have missed, your copy-editor can also act as a representative of your readers. Copy-editors, particularly those working in scientific, technical or medical publishing, usually have some training in or knowledge of your field, and will be able to spot ambiguities, inconsistencies, abbreviations that need explaining or text that doesn't match the pictures. If the copy-editor needs your advice on any changes, he or she will prepare a list of queries. These will either be dealt with by your editor, or will be sent to you before your typescript goes to the printers, in which case you should deal with them as swiftly as you can. If the queries are minor or few they will appear in the margins of your proofs.

There will be some changes which a copy-editor will make to your manuscript which may appear trivial or unnecessary. These will have been made to make your work conform to the publisher's 'house style'. You may have been sent a copy of the house style before you prepared your manuscript. These documents vary from one sheet of paper to a small book, and will contain the kind of instructions which are a nuisance for authors, though you should read them and follow them as far as you can, particularly with regard to the layout of tables, captions, references, lists and headings. However, the house style is an essential guide for copy-editors, who may work for several different publishing houses. It represents the choices your publisher has made on questions of style which could otherwise be answered correctly in a number of different ways. Do you number your lists 1 2 3, (i) (ii) (iii), (a) (b) (c) or (*a*) (*b*) (*c*)? Is 'Before Christ' written BC, BC, b.c. or bc? Is it energise or energize? Sixties, 60's or 60s? Syllabuses or syllabi? Bandgap, band-gap or band gap? These points may seem unimportant, since they have very little to do with whether or not what you've written makes sense, but publishers can be obsessively loyal to their house styles, and it is the copy-editor's job to change your manuscript to fit.

The copy-editor's other job is to prepare the manuscript for the printer. Apart from marking all the changes already mentioned, the copy-editor will instruct the printer on how to set the headings, text and tables, how to deal with mathematics and data, where to put pictures, and so on. Even if your manuscript has been printed out by a sophisticated word processor there will still be a lot of work for the copy-editor to do. This is why it is essential to double-space the lines throughout your manuscript – in the preface, notes, footnotes, references, captions and tables as well as in the text – so that there is enough room for the copy-editor's instructions to the printer.

By now, your book will be the responsibility of the publisher's production department, which will write a schedule for its progress to the presses. The copy-editor will be told a date by which the manuscript must be ready for typesetting, and by that time a typesetter will have been booked to prepare your proofs. The typesetter's relationship with your publisher is purely a business one: the typesetter has no interest in or influence over the content of your work, and will only change material which is impossible to set in the way you want. You will probably not know who your typesetters are, and if you do, there is no occasion on which you should deal with them directly.

On the nature of proofs

A Russian typesetter once mis-spelt Stalin's name. 'Stalin' means 'man of steel', but in print he appeared as 'Salin' – the man of lard. The typesetter was promptly executed. While publishers rarely resort to such drastic tactics, correcting errors is expensive, and so when you read your proofs you should limit yourself to putting right errors of fact or typography and to making last-minute updates. Questions of style and content should be dealt with before you submit the manuscript.

If the pages in your book are to be in the standard format – a single column of text which is the same width throughout – the first proofs you will see will be 'page proofs'. The text will have been broken up into pages, and the pages will have been numbered – 'paginated'. In any publication, whenever you open it, you will find that the page on the right will have an odd number. If your work is for a journal or is a single chapter for a book, the page proofs, which are usually printed with two pages to a sheet of paper, may have your page 1 on the left-hand side. However, the pagination will tell you which page is which: pages with odd numbers will be right-hand pages when printed. It is important to be clear about this when you read your proofs, because you want to be sure that the pictures or tables appear opposite or next to the relevant text. You may find you have to turn a page of your proofs to see the pictures, but you won't have to once the book is printed.

If your work is for a magazine or for a book with a complicated page layout, your proofs will be printed as 'galley proofs' or 'galleys'. These are continuous strips of text, with no page numbers, and may have tables, captions and sometimes equations grouped together by chapter or at the end. Once you've read the galley proofs they will be sent for correction and made into page proofs. The process of preparing a page layout from a galley proof is called 'pasting up'. Someone – possibly your editor, copy-editor, printer or a page designer – will cut up the galleys and paste them onto sheets of paper which are marked to show the dimensions and layout of the page. If the artwork or photographs are not ready by this time the person preparing the paste-up will have to guess the size and shape of the spaces they leave for the pictures, and you may find when you finally produce your photograph that it will have to be very small, or will have some vital detail sliced off one side so that it can fit. It is therefore to everyone's advantage if you can get as much of the artwork as you can ready before the paste-up is made. You will probably have an opportunity to see the paste-up before it goes back to the typesetters, who will use it as their guide to turn the galleys into page proofs.

Your editor should be able to let you know in advance when your proofs will be ready. Once the parcel of proofs lands on your desk you will have only a short time to read them, so make sure you don't leave for three weeks in the Bahamas the day before they are due to arrive.

There is a standard set of symbols which you should use for correcting proofs. They are a neat and clear way of making corrections, and you will be better understood if you use the printer's language. Proof correction symbols come in two parts: you use one part to show where the correction goes in the text, and the other, which goes in the margin, to say what the correction is. If there are two corrections of the same type in one line, make two marks. Your editor may be able to provide you with a list of these symbols, but the most useful ones look like this.

Delete unnecessary characterss or words words.

Insert mising characters or words

Replace a charactor or verb. words

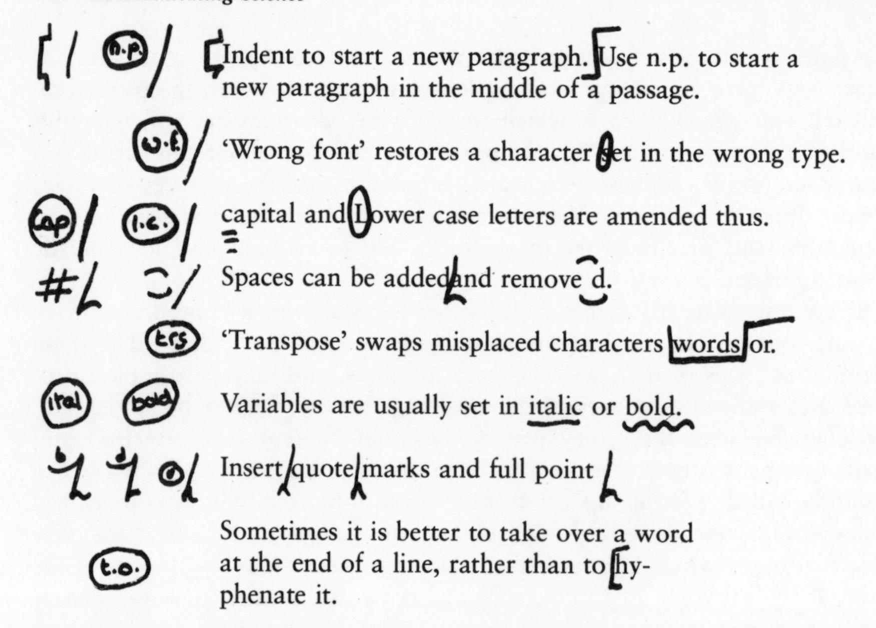

You should draw a bubble around anything you write on a proof which is an instruction, rather than material which needs to be printed.

The purpose of proof-reading is to correct errors, not to make stylistic adjustments or to add or delete large amounts of text. Deleting chunks of text at proof stage causes as much disruption as the sudden appearance of new paragraphs. Try and make your corrected text fit into the space occupied by the text you are correcting. Make corrections that are the minimum required: if you are changing one letter of one word you don't have to write out the entire sentence. Although in order to correct your one letter the typesetter may reset the whole sentence, it will be clearer if you mark the minimum possible. The person responsible for the error pays for the correction, so mark your own errors in blue (for the publisher's costs) and typesetting errors in red.

Ultimately, the author is going to have take responsibility for everything that appears in the book – ideas and typographical errors. Your name is going on the cover, and if there's anything you don't like when it's published you have no right to blame your publisher. Although a professional proofreader will see your proofs, you will see them first. Do as much as you can. It is sometimes difficult to see the mistakes when you are familiar with the text – you see what you expect to see – so ask a friend to help. Your proofreader may have some competence in your subject, but you are the expert; take care with maths, technical terms, data and proper names. Read everything: contents, preface, index, footnotes, references, headings . . .

Writing an index

If you can divide the number of pages in your book by 64 and still have a whole number, your publisher and printer will be very pleased indeed. They may be content if you can get a whole number by dividing by 32 or 16, or, at a pinch, 8. This has nothing to do with any mathematical predilection, but is a consequence of the way books are printed. Each batch of 64 pages starts life as a single sheet of paper. 32 pages are printed onto each side of the sheet, and then it is folded in half, and in half, and in half . . . until, by some miracle of three dimensional geometry, the pages are in the right order.

Your editor will have decided how long your book will be. Once the text has been typeset you will be able to tell how much room there is left for titles, the contents and the index. If your book is to be 128 pages long and the text takes 122 pages, you will find that you have two pages for a title, two pages for a contents list and two pages for an index. If you have written an index that runs to two and a half pages, in order to print it the publisher would have to add four more pages to the book: because of the way the pages are held together, four is the smallest number that can be added. Given the choice between adding four extra pages and cutting your index, the editor will almost certainly cut your index. This means that if you are told that you have two pages for an index, then there is no chance at all that you can have two and a half. The number of entries that will fit into two pages will depend on the type size and layout, so ask your editor how long the index can be before you start.

Indexes are a chore to compile, but are a very important part of any book, and particularly of a book that contains a lot of detail or technical information. Indexes are also a sort of abstract or summary – they represent what the indexer feels are the important points of the book. Browsers in bookshops can be seen looking at indexes: if the index of your book on astronomy shows a number of references to Kepler, Johannes and only one to NASA, the would-be astronaut will probably look elsewhere. This is why it is important for authors to write their own indexes, for authors know better than anyone what their book is about. If you can't do the index yourself, bribe a colleague who knows something about your subject to do it for you. You could hire a professional indexer through one of the indexer's societies – they keep registers of indexers who are experts in a particular subject. The societies' addresses are given on page 173.

The best time to prepare your index is when you read the page proofs. By this time all the text will be on the pages on which it will be published, and the pages will be numbered. The publisher will provide you with a spare set of proofs to use for preparing your index. Ideally you should read the index proofs and the proofs you are correcting separately, but if you are pressed for time you can deal with them together. As you read and correct your proofs, mark on the spare copy where there is information which you want indexed. Then go through the spare proofs and make a list of what you want to put as the index entry, together with the page number. A word processor is useful at this point, as it can put the list in alphabetical order for you. Otherwise you'll have to write each entry on a separate piece of paper, put those in alphabetical order and then type up the list. If you can put your final list onto disk it can probably be typeset directly, which will save a lot of tedious checking later. Keep the proof with the index marks on it, as then it will be much easier to amend the index should you have to change page numbers.

Some things need not appear in the index. These include anything which is mentioned in the contents lists or chapter headings, and occasions when something is mentioned although no information is given about it. You do not need to index your preface or introduction, unless there's material there which crops up nowhere else in the book (in which case, why is it in the introduction?), and don't index the references or bibliography.

Recommended reading

Judith Butcher, 1980, *Typescripts, Proofs and Indexes* (Cambridge: Cambridge University Press).

The Chicago Manual of Style, thirteenth edition, 1983 (Chicago: Chicago University Press). The standard reference for editors, it offers a straightforward guide to all stages of book production. This edition includes a section on electronic publishing.

Arthur Plotnik, 1982, *The Elements of Editing: A Modern Guide for Editors and Journalists* (New York: Collier). A succinct and realistic 'insider's look' at the relationship between writer and editor.

Some publishers produce a 'guide for authors'. If yours doesn't, another publisher will be able to sell you one. Basil Blackwell's is good: its addresses are on page 173.

III. *Speaking*

Speaking

You might not want to write, and you may be able to dodge the attentions of a radio interviewer or television reporter, but sooner or later the day will come when you'll be invited to talk about your work to a potential sponsor or to speak on the delights of a career in science at a local school – and you may not be able to refuse. Twenty years ago, the chances of a scientist being invited to speak anywhere other than at an adult education institute or the local natural history society were slim indeed, but today schools, clubs, community centres, charities and pressure groups all welcome a scientist who is willing to talk in an interesting way about subjects ranging from space travel to the silicon chip, from plastics to pig-breeding.

There was a time when a public speaker could get away with a lot of meaningless waffle and rhetoric. Not any more: with television and radio in every home, people are now expert listeners, and have become used to hearing words which inform and entertain. They can no longer be fobbed off with empty phrases and careless delivery – not even from scientists.

Saying it

> The man of science appears to be the only man who has something to say, just now – and the only man who does not know how to say it.
>
> J.M. Barrie, dramatist and journalist

Journalists often make great efforts to find the human interest angle for a science story. Spoken presentations have the human interest built in – you, the speaker. Speaking is a very personal form of communication: unlike writing or broadcasting, you are communicating directly with people who are right in front of you. You will be able to see and hear their immediate reaction to your efforts. Unlike authors who never know whether their readers turned to the sports pages after the first paragraph, speakers can see if their audience leaves early or falls asleep. This is a daunting prospect, but is no reason to worry: you know that the people who have invited you consider you well-qualified for the job, and that the audience, who have chosen to come to hear you, are interested in what you've got to say.

Some people think that public speaking is a mixture of acting, salesmanship and shouting. It is a ritual of a kind, but your task is simply to find an effective way of conveying information and impressions. Public speaking can be thought of as a kind of enlarged conversation, but enlarged only in the sense that you will be addressing greater numbers of people. It is far closer to the talking that might go on around a dinner table than to what happens on a theatrical stage. But the kind of conversation that might just pass among friends is liable to drive a less committed audience to distraction.

'When are you leaving?'
'Tomorrow.'
'Tomorrow?'
'Yes, at about 12 o'clock.'
'Um, I wonder what the weather'll be like . . .'
'Pretty good I suppose.'

And so on. This kind of exchange has all the noise but not enough of the content and form of a good public talk. In the first place, when you give a talk you can speak to your audience without interruption, which allows you to organize and plan ahead. This relative freedom carries the responsibility of delivering your talk in a coherent and entertaining way. If people cannot interrupt you, they can all the more easily switch off, doze off, or even wander off.

The next difference between public speaking and a conversation is that public speaking is conversation with a purpose. You talk to an audience for a specific reason – you want them to feel, think, or do something – and this objective must be central to your work. It is the essence of successful public speaking: know what you want to say, organize your talk to say it. But there is far more to a talk than simply the words: you have to think of stage management, planning, conquering nerves, and coping with emergencies, not to mention a host of practical skills ranging from voice production to drawing on the blackboard. You might think that anything else would be mere embellishment – liveliness, commitment, enthusiasm, vitality . . . – but the reason why these qualities are not optional extras is that without them, your audience may well not stay awake long enough to get the message.

When you give your talk you are not only the author, but also the editor and publisher – the writer, performer and director. You have a great responsibility, and only yourself to blame if things go wrong. You can also take all the credit for your successes, and the thrill of the moment when you realize that the audience is with you and responding just as you had hoped will more than make up for all the effort and worry.

Most of what we've said about writing – about words, style and form – also applies to speaking. In this section we look at the skills that can help a scientist who takes to the stage. The section is in three parts: the first deals with preparation, the second with performance, and the third with visual aids.

3.1 Sit and Prepare

When you are invited to give a talk, ask a few questions before you accept. First, ask yourself whether you are qualified to give the talk – and 'qualified' means enthusiastic about the subject, as well as knowledgeable. Then ask the organizers why they chose you. Were you recommended, or was your name picked out of a hat? Ask about the occasion and, if you don't know them, ask about the organizers. Then consider your own schedule: where and when is the talk? How will you travel? Will your expenses be paid? Will you need to stay overnight, and if so, where? Who is your contact in case of breakdown on the motorway? What arrangements, if any, are there for 'hospitality' before or after your talk?

If the arrangements, the topic and the date are satisfactory and you decide to accept, you will also need to find out whether anyone else has been invited or whether you will have the platform to yourself, and how well any previous talks have been received. This will give you an idea of what is expected of you. How large will the audience be? Do they represent a particular group – athletes, accountants, amateur anthropologists? How old are they? They may not remember the sixties, let alone the last war. How much do they know about your field? Why will they have come? What is their attitude likely to be towards you, and to your material? Is there any difficult area in your subject, where you might need to be particularly sensitive to the audience's beliefs or background?

You will also need to know the boundaries of the subject of your talk – you may find after straying into an interesting side issue that you have repeated a previous speaker's speech. What is the purpose of your talk? Are you supposed to be teaching your subject in a rigorous and serious manner, or are you the light relief? How long should your talk last? If you are booked for an hour and feel you can't last more than forty minutes, say so.

Woodrow Wilson was once asked at a press conference how long he took to prepare a ten-minute speech. He replied 'Two weeks'. The journalists were surprised, and asked 'How long for a one-hour speech?' Wilson replied 'One week'. When the journalists asked how long for a two-hour speech, Wilson said 'I'm ready now!' Be warned: it is much more difficult to prepare a short talk than a long one. If you had three days to explain thermodynamics or ecology you could start now and meander through the subject in whatever order, repeat yourself and backtrack and still get your message across. If you only had fifteen minutes you would have to think very hard before you started or you would achieve nothing. Don't think that a short talk will need less preparation than a long one – it will probably need more.

Planning a talk

Accepting invitations to speak is easy; fulfilling them is more difficult. First, ask yourself the four questions on pages 33-34, which apply to speaking as well as to writing.

Why should anyone be interested in what I've got to say?
Why is my subject important?
Why am I doing this now?
Why should anyone trust me?

Once you are sure that you can answer these questions satisfactorily, you can make things much easier for yourself by beginning work on your presentation straight away. What kind of talk should you give? This will depend on the occasion, the subject and the audience. Different audiences will value different arguments: if an audience is interested in the economic potential of your work they may not be impressed if you concentrate instead on the elegance of your method. What will be your main motive when you speak? Are you trying to entertain, convince, motivate or simply inform? Your approach should be chosen accordingly. You will find that once you have decided what your talk will be like you will work almost subconsciously – ideas and examples will tumble into your mind while you are busy doing other things. Carry around a notebook and write down your ideas as they crop up. A newspaper story, radio broadcast, or even some remarks overheard in a bus queue could provide you with appropriate material.

Planning is even more important in speaking than in writing – the reader can always read a piece of writing again, whereas listeners will hear the talk only once, and won't be able to go back over difficult parts or skip the boring bits. This means that the talk must be logically organized and the information paced so that you make sense and maintain the audience's interest. It can help if you plan a couple of points where someone who has lost track of the argument can find their way back.

The first section of any talk is, by definition, the introduction.

Introduction

A strong introduction is vital: if you lose the audience's attention right at the start you'll have to do something extraordinary – like spontaneously combust – to get it back. However fascinating the topic may seem to you, the audience may not at first take the same view, so start with a question, a startling statement, an anecdote or a challenge. The same general points about bright, lively openings in writing apply to speaking as well, although when you speak you can afford to take your time – it wouldn't be wise to tell your whole story, newspaper style, in your first paragraph. You may start from the known and then work towards the unknown, or perhaps connect what you have to say to a current event – a recent headline can give a talk contemporary relevance. Knowing about public attitudes could get you off to a good start: 'more than 40% of people recently asked said that you'd have to be a raving lunatic to go to see a psychiatrist'.

What happens in the main body of your talk will depend on what subject you have chosen. However, a general outline might be as follows.

My talk is about . . .

Explain precisely what it is you are going to talk about. Define new words and concepts; discuss history, background and personalities.

To set the scene . . .

Discuss the current situation: successes, problems and aims.

These are my ideas . . .

It is important to give your ideas before you give your evidence, so that when the audience sees the evidence they will know what to look for.

This is my evidence . . .

Present your evidence in such a way that it illuminates and affirms your ideas.

These are my conclusions . . .

Make sure that your conclusions are based entirely on material you've presented during your talk, or the audience will think that your reasoning is flawed.

And to finish . . .

If first impressions are the most striking, last impressions are often the most lasting. While it is quite in order to startle your audience to attention at the start of your talk, you need to exercise more caution in drawing to a conclusion. At the end of his play *Man and Superman*, Shaw gives the stage direction: 'Go on talking'. The speaker would be foolish to follow that advice – many a fine talk has been spoiled by a conclusion that trails off into the sunset. Shakespeare's advice isn't much better: if you 'stand not upon the order of your going, but go at once', you will leave your audience shocked by the sudden silence.

One useful way of concluding is to summarize what you have said: this gives the audience a chance to order their thoughts and may prompt them to ask some questions. It is a particularly useful strategy when you are dealing with a lot of information or with very technical material. If the audience understands your summary they will go away thinking that they understood the talk, even if they didn't. You may, however, be discussing matters that are less concrete – the life of a botanist, perhaps, or 'why I decided to become a mathematician', in which case it might work better to state once more your main theme. Let the audience know when you are nearing the end of your talk: say 'to finish off, I'd like to remind you of the purpose of this work', or 'just to sum up, the most important implications of this research are . . .'. Don't say 'finally', 'to conclude', 'lastly' or anything similar unless you are actually about to stop. Once you've told the audience that you are about to finish, make sure you do so within a few minutes.

A plan in action

General suggestions are all very well, but how would they work in practice? Let us imagine that you are a computer scientist working on a program to analyse the mechanisms of the heart beat, and you've been asked by a local adult education institute to talk to the students about your research. This is the kind of invitation you should accept: it offers an audience which enjoys learning and little inconvenience of travel and time. The problem is that your work is in computer modelling, devising programs to simulate the heart beat and testing the results against experiment, which is very technical. The first thing you must do is to see where you can develop links between your own research and the interests of your audience, and to do this you need to know who the audience are. In this case you can guess that they are well-educated, interested adults who know that heart disease is a major cause of premature death, and is related to diet and exercise. While your audience may not yet be interested in computer modelling, you can use their interest in heart disease to introduce them to your field.

You might decide to tell the audience about some recent results of your research, to show them how it might be important in solving real problems in health care, and along the way to give an insight into how scientists go about their work. This means that your listeners will acquire some knowledge of science as well as some scientific knowledge. In the best traditions of informal learning, you will offer them 'rational entertainment'.

One of the first things you will be asked for is a title. Try to make your title inviting and accessible, but also as direct as possible. 'Forty Years in the Fume Cupboard' is intriguing, but not as informative as 'My Life as a Chemist'. 'Endoparasites in Humans' is plain enough, except that it scores two out of three for jargon. You would attract a much bigger audience with 'The Bugs on Your Body'. You could try 'Mending Broken Hearts', or 'Computer Care for Healthy Hearts'.

Introduction

George Bernard Shaw once said that he avoided putting anything of any importance into the first few minutes of his plays. Audiences were invariably late, and even if they were on time they took five minutes to change seats, settle down and open boxes of sweets. When you give a talk, all this preliminary shuffling and crunching will be done while your host introduces you. When you start you can launch straight into your talk. You simply need to use the first few moments to catch the audience's attention.

You could tell the audience what they already know, and then challenge that knowledge:

> Some people think that heart disease strikes only the old . . . Some think that if they have heart problems, a transplant is the only answer . . . Others think the main causes of heart disease are fatty food and smoking . . . But the only fact of the matter is that we know very little about how the heart works. Today, I want to tell you about my work on what keeps the heart beating. Once we understand that, we'll be a little closer to helping people with heart problems . . .

A controversial statement also works very well, though it will take some self-confidence to put it across. If you begin like this, your audience would soon stop shuffling in their seats and give you their undivided attention:

> Heart disease is the biggest killer in the world. One out of three of you here today will die of heart disease. That's an awful statistic, but scientists around the world are doing their best to understand why heart problems are such a killer. Part of the answer lies in the way the heart works, and to understand that we need to understand the heartbeat.

You could try using a prop:

> This tiny piece of plastic could one day could make many people a lot healthier and happier . . . No, it's not the entry pass to Fort Knox: it's a computer disk that carries information about how the heart beats . . .

There are many different styles of opening a talk, but to establish a good atmosphere you always need to win your audience's attention and interest. Having achieved that, your listeners are on your side.

My talk is about . . .

You cannot cover every aspect of heart disease, so this is the moment when you should home in on your particular topic. In a concise, clear and straightforward statement, tell the audience what your talk is about, and how you are involved.

> Today I want to talk to you about what we scientists working behind the scenes at the Hospital have been doing to try to help people with heart disease.
>
> I'm going to talk about how our heart keeps beating, and about how my work with computers can help us understand this process.

Setting the scene . . .

Now is the time to tell them what is so interesting about your topic, and why you have selected it. Develop your ideas perhaps by showing the broad application of your work, its background, its relationships with other problems, or why it is important. As you do so you could say something about how you fit into this picture yourself. Your audience will be interested to learn how your ideas arose, and why you have the interests you do. What do scientists actually *do* all the day long? People are ignorant of but interested in what goes on behind laboratory doors. The more esoteric your research is, the more important it is to present it with a human face.

> I first became interested in heart disease about eight years ago. At the time I was working in computing, and out of the blue I was offered a job in a medical engineering department at the Hospital. I wasn't too sure what 'medical engineering' was, but I decided to give it a go. They'd suddenly got some government money to start a new project using computers to analyse disease. They needed a team of computer scientists to work alongside the medical researchers. Now I write computer programs which imitate the action of the heart, and then we try to work out what effects different things, such as diet, stress, general health and chemistry, have on the beating of the heart.

You have now told your listeners who you are and what you do. You have also begun to destroy some of the myths people believe about scientists. You have told them that you work in a team ('they needed a team of computer scientists'). Your career suddenly changed direction – a common experience for scientists – and you stepped into a new field and took up a new challenge ('I wasn't too sure what medical engineering was, but I decided to give it a go'). Scientists' careers and the progress of research depend on what funds are available ('they'd suddenly got some government money for a new project'). Different disciplines overlap in research (' . . . computer scientists . . . work alongside medical researchers . . . '). By including this personal note, you have told your audience something about yourself, and a lot about science. Anything you can do to create a more accurate impression of what you do is helpful in creating a more realistic public attitude towards scientists' work. Even if 90% of the content of your talk is forgotten after a week, the impression you create will last.

You should now disclose your point of view as neatly and explicitly as possible.

These are my ideas . . .

Some speakers feel bashful about presenting ideas of their own directly. Don't worry about this – after all, your idea is a considered judgement which you will shortly develop and substantiate, and there is nothing audiences like more than having a clear and succinct statement of a speaker's point of view. You were invited because they want to know what you think.

> I think that far too little consideration has been given to how the heart actually works.

> I believe that we can come closer to a solution to the problem of heart disease if we study the heartbeat.

You might pose questions and provide answers, or be provocative:

> The beat of the heart is like the tick of a watch: when it stops, everything comes to a sudden halt. A lot has been said recently about the importance of diet and smoking, but to my mind that's not enough – we need to understand more about the physiology of the heart. Without that knowledge we are like a watchmender who has never seen a mainspring.

Why is stating your ideas so important? Science is a personal activity: what we know and believe of scientific matters comes from scientists – people – whose judgement we trust. If you tell your audience 'it is widely accepted that', or 'according to the conclusions of the Oxford group', your audience won't know whose judgement they are being asked to accept, and will have no grounds for either challenging or accepting it. But when you stand up before a room full of people, introduce yourself, talk about your work and give your opinion, the audience knows exactly what it's getting.

The audience is now ready to listen to your main points.

This is my evidence . . .

This is where your detailed knowledge of the subject comes in, and where preparation is essential. Describe how the heart works before you describe how it doesn't. Say what a computer does, and then say what you do with the computer. Don't overload this section: people are able to take in only a small amount of information over a short period of time. As you present and develop your main thesis, pay particular attention to the order in which your overall points are made, and to how you move from one point to another.

These are my conclusions . . .

Your points should have been presented in such a way that your audience arrives at your conclusion when you do. Once you have explained how your conclusion arose, state it in a single sentence:

> This means that if we put information about diet, lifestyle and medical history on to this disk, we can use the program to predict the consequences for a person's heartbeat, and thereby help them to keep a healthy heart.

And to finish . . .

You might want to end by stating your main ideas again, or even by coming back to your opening remarks.

> Well, we have covered a lot of ground in the last forty minutes. I began this evening by saying that I had a piece of plastic which might eventually help to make many people happier and healthier. I hope that you can now see what I meant.

You could end on a personal note:

> Eight years ago I was an ambitious and eager computer enthusiast, waiting for a worthwhile job. If you had told me then that I would spend eight years working on the physiology of the heart I would have laughed. Science is certainly full of surprises!

At this point we can say goodnight to our computer scientist. The plan would work well for reporting the results of scientific work, but not so well for a review or history. Instead of a logical sequence, think of a time sequence, or a geographical trail. Maybe your story relies on a series of personalities, or is incidental to a sequence of political or social events. Whatever it is, there must be a thread running through the talk, and this provides the basis of your plan. If you have no thread, you don't have a coherent talk. If you have more than one thread, find ways of crossing and linking them, and if you can't, sacrifice all but one of them and concentrate on a single theme.

Making notes

Having brought together ideas and information, thought hard about the kinds of things likely to appeal to your audience, arranged your material in an appropriate order and written a plan, you are ready to begin writing your talk. By now, you will have realized that the most demanding part of giving a talk is not getting to your feet to deliver it, but sitting down to prepare it. This is partly because you will have too many ideas buzzing around in your head, and partly because the priorities of a general audience are not necessarily those you would choose for a scientific audience. Whatever you do, and however you decide to organize your plan, remember your audience at all times.

You may find that you have too much material for the time allowed for your talk. Put your main points on cards or separate slips of paper, then spread them out. This process will clarify your objectives, and the process of elimination can begin. How does each part fit with the particular approach you wish to make to your general theme? And how is it adapted to the particular audience you are to address? What order do you have in mind? How is each point illustrated, explained or developed by the information you intend to present? Could you have used different examples?

As you organize your material you will find that your ideas and information settle into an order. There will be some information which your audience must have, some which they should have, and some which is not essential but which you think they would find interesting. You will find that sorting out the information in this way will help you to organize your own thinking.

If you are gifted with great clarity of mind, and are tidy to boot, you may find that all you need as a memory aid is a rough plan, however much it may be covered in scribbles and crossings-out. It was said of the seventeenth-century Frenchman Bishop Bossuet that he could preach off the top of his head on any text which was placed before him on the pulpit cushion. One Sunday, the paper on the cushion was blank. Holding it up, he exclaimed, 'Here is nothing – nothing at all! Yet of nothing God created all things!' and proceeded to preach upon the Creation. It is unlikely that you will be able to make as much ado about nothing as the Bishop, so what alternatives are there for you once you've prepared the substance of your talk?

If you are presenting material which is unfamiliar to you (perhaps you had only a week to prepare for a public inquiry), or it is important that you get the details exactly right (you are giving forensic evidence in court), you can read your talk word for word from a prepared text. This has the advantage of calming your nerves and will certainly get you through any talk without too many hitches. Max Planck used to read lectures word for word not from a typescript but from his published books, no doubt saving students the trouble of listening to him at all. Perhaps quantum theory was bold and startling enough to keep his audience awake.

The real drawback of Planck's method of reading from a text is that it is dull. Unless you are an expert at reading from scripts, you will find that you look only at the paper, and never at the audience. You voice will lose all expression. You will sound quiet, and what the audience can still hear will be written language, which is very difficult to listen to. Your personality and spontaneity ratings will be zero. If you feel that you simply must have every word written down in front of you, try talking from notes into a tape-recorder, and then writing down your spoken words from the tape. However, a pile of typed paper always dismays an audience – the pile never, but never, seems to get any smaller.

While marriage to your script is not ideal, divorce isn't recommended either. Learning your speech word for word and then abandoning the script is very risky. If you can deliver it with the skill and style of an actor you may be a raging success, though some may suspect that you're giving the 101st performance of your party piece. You will find it difficult to make last minute changes to your script, and will be thrown by any interruptions. If you are distracted and forget your lines, which happens to the best performers on stage, you won't hear a whisper from the prompt in the wings, but will be alone and silent, facing a sea of expectant faces.

The best method of keeping going when you talk is a combination of these: have your words partly written down, and partly in your mind. What you write down, in the form of key sentences or short notes, should be enough to remind you of the material you have prepared. One way to compile these notes is to start with your basic plan and work up to a more detailed one. Another way is to write out the complete text of your talk and then go through it underlining key words and phrases (remembering that adjectives and verbs can jog the memory as well as nouns). Then make a list of what you've underlined, and put the full text away. You may find it helpful to include in your notes the first sentence of each paragraph. If you find that your notes consist of long paragraphs you are probably not familiar enough with your material. Quotations and data should be included in full.

Spoken language is different from written language, particularly in science. This is easily demonstrated – try reading a paper from *J. molec. cell Immunol.* or *Proc. natn. Acad. Sci. USA* out loud. You may find that when you talk about science you unwittingly slip into a 'scientific' style of speaking. This will be difficult for a general audience, so try to talk in a conversational style. This doesn't mean adopting what you presume to be the audience's slang – you are bound to get it wrong and will sound patronizing. There are phrases which you might write but would never say – don't let them sneak into your notes, or you may find yourself saying them.

Numbers should be simple – say 'one in twelve' rather that '8.3%', and visual representations are better than mathematical descriptions. Hand waving has a poor reputation in scientific circles, but can be used to good effect. 'It fluctuates sinusoidally' carries none of the explanatory power of 'it goes like this' accompanied by an undulating arm.

Check the pronunciation of foreign or difficult words, and practise saying them. If you can't find the standard pronunciation, or you simply can't pronounce a word correctly, give your own version confidently and without apology. Use English rather than foreign words and idioms wherever possible – if you find yourself translating a phrase into English, cut the phrase and just give the translation. Use the English plurals of foreign words – say 'plateaus' and 'formulas'.

Phrases such as 'this bit is irrelevant, but I thought I'd mention it' and 'you can leave out this step if you prefer' are irritating for an audience: if something is irrelevant, don't mention it. Similarly, 'I'll explain this in a minute' and 'we'll come back to this later' are confusing and evidence of bad planning. Don't tell people what you are not going to tell them: if you say 'I'm not going to discuss the environmental aspects of this process', the audience will wonder what it is you're not telling them – and why not. If you are not going to discuss the environmental aspects, then don't mention them. Never say 'I won't go into that because it's too complicated'; if you do, the audience is entitled to conclude that you think they are stupid. Decide before you start writing your notes what information will be appropriate for your audience, and stick to that.

A scientific talk is not the occasion for a stand-up comedy act, nor for anything that could be described as telling jokes. The reason is simple: if that was what your audience was looking for, they'd have gone to see a comedian. Nevertheless, laughter makes a valuable contribution to any talk. Like applause, laughter turns a passive, silent audience into active participants in a shared experience. There are a number of books which will supply you with off-the-peg anecdotes, but they are rarely a good fit. Instead of digging up someone else's joke, collect amusing incidents from your own experience. Ask around at work – someone may have just the story you need.

Practice makes . . . better

The only way to check whether your notes are complete is to have a practice run. Deliver the talk using only your notes to prompt you, and amend them at any point where you get stuck. Then tape yourself giving the talk, or get a friend to listen to you, and amend any points which are ambiguous or repetitive, or where the argument jumps. Make sure that the highlights are spread throughout the talk, rather than crammed together at the beginning or end. Mark in your notes where you are going to show a slide or refer to a model. Give yourself an idea of time – mark the notes to indicate when you are half-way through, or write what the actual time should be when you reach a certain point. Write incidental notes about visual aids and timing in a different colour, or you may find that you read them out.

Time your practice runs to see if you are likely to finish on time. Then try reading from a book at a rate of 120 words per minute, and listen to the speed. It may sound very slow, but it's roughly what people can assimilate when listening to unfamiliar material, particularly in a large auditorium or one with poor acoustics. Trying to achieve 120 words per minute when you give your talk is an unnecessary complication, but keeping the sound of that speed in mind while you practise will help you moderate your pace, and you'll have a better idea of how long your talk will last on the day.

It is important that your talk should last for the specified time. Finishing early is not such a problem, unless it was forty minutes early in a one-hour talk, in which case the audience is entitled to feel cheated. If you do find yourself ahead of schedule it may be because you are rushing, in which case you should slow down a little. Going over time is a crime which is rarely forgiven, particularly if other speakers are to follow you. The audience won't be too pleased either – they will feel bound to stay simply because they are too embarrassed to leave, even if they do have a train to catch. They will shuffle in their seats, pack up their possessions, take their jackets off the backs of the chairs and look at each other with expressions which read 'shouldn't she have finished twenty minutes ago?' They will not notice your magnificent conclusion because they will be laying odds on what time you'll stop.

If you find when you practise that you have to shorten your talk, make fewer points. This will be more effective than cramming the same number of points into less time – the audience can only absorb a certain amount of unfamiliar information over a short time, and will need some padding to cushion the new material. Since you may find on the day that your talk runs for longer than in practice, include a section towards the end which you can cut without ruining the structure of your talk. Write this section in your notes under the heading 'If it is now 6.50 or later, miss this next bit out.' (This method only works if you remember to wear your watch . . .) Be particularly familiar with your concluding paragraph so that you can, if necessary, deliver it early, so that you can finish neatly rather than stopping in mid-stream. The audience probably won't notice that you've missed something out, but if they do, it will give them something to ask about afterwards. If they do point out the gaps, there is no need to mention that you left the material out deliberately.

Usually when people practise their talk they go through it much more quickly than they would on stage. They mumble their words, flick through their slides and hope for the best. This sort of practice is not very valuable – you have to give the talk for real. Walk across your room and pretend to write on an imaginary blackboard if you intend to use a blackboard in your talk. Writing on boards, changing overhead transparencies and having a drink of water take up a surprisingly long time. Deliver your main points and conclusions, and introduce new words and ideas, slowly and deliberately. It won't seem too slow to the audience, who will need time to assimilate the new material. The phrase 'making your point' derives from an old theatrical custom. 'Points' were moments of high drama, where the performer would break from the action, come down to the front of the stage and speak the lines directly to the audience. Such points were emphasized further by the audience, who were encouraged to respond vigorously, either with applause or with rotten fruit. While you don't have to follow the theatrical tradition to the letter – and neither does the audience – make sure that everyone knows when you are making an important point. Pause, make your statement slowly and emphatically, and then pause again.

Pauses are very useful, both for you and for the audience. A pause indicates that you have finished one thought and are about to begin another – it is the spoken equivalent of starting a new paragraph. Pauses always feel longer to the speaker than to the listeners – what feels like an eternity may be only three or four seconds, but this will give listeners time to collect their thoughts and you a chance to take a deep breath and sort out what you are going to say next.

Having a first-rate set of notes does not guarantee a first-rate talk. Practice may not make you perfect, but it will definitely improve your performance and should not be under-rated. Give your talk from your notes several times before the real thing – to a tape-recorder, a mirror, a video camera, and to your family, colleagues and students. Practice is only useful if you learn from it, so be critical, and encourage your guinea-pig audience to criticize you. Don't let criticism dent your self-confidence – being able to improve as a result of criticism should make you more confident.

Once the notes are complete they can be written out for use on the day. Type them or write them clearly on one side only of file cards or small sheets of paper. Tie the cards together so that they stay in the correct order, and if you use paper put in a ring binder rather than stapling it. That way you can turn the page unobtrusively and without making rustling noises, which are a particular problem if you are using a microphone. Number your cards or pages, and keep an eye on the numbers so that you don't turn over two at a time.

Public speaking is very flexible: you have no editor or producer to please, and a wide range of ways of reaching your audience. You could be speaking in a prison or on a cruise liner, to trainee astronauts or children. How you prepare and deliver your talk depends on the time you have, your audience, what you have to say and how experienced you are. Choose your method to suit the circumstances – yours and the audience's. Talking to the public is tougher than talking to scientists: their standards are high. If your notes are good and you have prepared your talk thoroughly, there is no reason why time or your audience should have cause to run out on you. All you have to do now is stand up and talk.

3.2 Stand and Deliver

> The appearance of a person so afflicted is characteristic. The face is strained with a furrowed brow, the posture is tense, the body is restless and often tremulous. The skin looks pale and sweating is common, especially from the hands, feet and axillae. Readiness to tears, which may at first sight suggest depression, reflects a generally apprehensive state . . . Symptoms related to the gastro-intestinal tract include dry mouth, difficulty in swallowing, epigastric discomfort, excessive wind caused by aerophagy, borborygmy and frequent or loose motions. Common respiratory symptoms include a feeling of constriction in the chest, difficulty in inhaling, and overbreathing and its consequences. Cardio-vascular symptoms include palpitations, a feeling of discomfort or pain in the heart, awareness of missed beats and throbbing in the neck. Complaints related to the functions of the central nervous system include tinnitus, blurring of vision, prickling sensations and dizziness (which is not rotational).[1]

You are bound to suffer from some of these alarming symptoms when you talk to an audience, for they are the symptoms of stage fright.

Conquering nerves

The fear of public speaking is more widespread than any other. We are all frightened of making fools of ourselves, of losing the attention of an audience, of being misunderstood or inaudible, of drying up or breaking down. For most people, nerves are the biggest barrier between them and an audience, and they often ask how they can rid themselves of stage fright, as if it were a noxious disease. Actually – and despite the description above – it is nothing of the sort. Nervousness is perfectly normal in the face of a situation that will test you and during which you will be judged. It is a throwback to life thousands of years ago when meeting an adversary meant you had to fight or fly. The rush of adrenalin through your body and the increased heart rate and blood pressure are natural responses to a taxing and exciting situation.

If you do feel tense at the thought of speaking in public, it may help to know that you are not alone. Even professional speakers suffer from stage fright, though on stage they appear to be at ease because they have learnt to deal with their fear. A brash and over-confident young actress, leaving the stage after saying her two lines, once saw the great Sarah Bernhardt shaking in the wings. 'Are you nervous, Madame?' the girl asked. 'Me, I never suffer from nerves on stage.' 'My child', replied Bernhardt, 'but you will when you have some talent.' Nerves may signal talent: they certainly mean that you are anxious to do well and are aware of your own fallibility. This acknowledgement is one of the strongest assets in becoming a successful public speaker, since recognizing your faults is the first step towards correcting them.

Stage fright cannot be completely eliminated or conquered, but it can be controlled. A good way to stop nerves interfering with your performance is to concentrate on what you are saying. The physicist Richard Feynman recalled his first talk with a mixture of horror and amazement. He was terrified – Einstein and Pauli were in the audience – and yet the talk went well. He later described the experience in his autobiography:

> Then the time came to give the talk, and here are these *monster minds* in front of me, waiting! My first technical talk, and I have this audience! . . . they would put me through the wringer! I remember very clearly seeing my hands shaking as they were pulling out my notes from a brown envelope. But then a miracle occurred, as it has occurred again and again in my life, and it's lucky for me: the moment I start to talk about physics, and have to concentrate on what I'm explaining, nothing else occupies my mind – I'm completely immune to being nervous. So after I started to go, I just didn't know who was in the room. I was only explaining this idea, that's all.[2]

If you can think about your talk as 'only explaining this idea', and can concentrate on the idea and the explanation, you will find that you can begin controlling your nerves. You may not become as flamboyant as Feynman, but in your own way you could easily become as lucid and appealing. What Feynman was able to do was to carry his own natural exuberance and enthusiasm into the lecture room, and to turn what might have been an inhibiting nervousness into fuel to fire his performance.

The best way to control your nerves is to be well-prepared. Your notes are complete, your practice runs went well, and all the arrangements have gone according to plan. Tell yourself that you are well qualified, well equipped, and the right person for the job. You can afford to be confident.

You will feel more comfortable when you speak if you are familiar with the auditorium. Arrive in good time to walk all round it, and sit in the front and back rows to see what view your audience will have of you. If you have any walking to do during your talk, perhaps to the light switch or to a model, make the journey a couple of times so that you discover any steps or protruding furniture before they catch you unawares. Check that your chair is comfortable and the right height, and see whether you'll be provided with a table or lectern. If you find you are expected to stand and you would rather sit, tell the organizer. Do not be bashful about stating your needs – good organizers, like good waiters, like to know as clearly as possible what is expected of them. Where is the pointer, water, chalk and light switch? Can anyone else can work the lights for you? Check the overhead projector and the slide projector. Meet the projectionist.

If there is a microphone, test it. See how the sound varies as you lean back or drop your head. Test the range – if you walk to the blackboard the audience may be unable to hear you. By speaking a few lines into the empty room, you should be able to work out whether or not you need a microphone, but remember that a room full of people will absorb more sound than an empty one. When you test the acoustics don't follow a 'one, two, three' or 'Mary had a little lamb' routine; familiar words will sound much clearer and so give a false impression. It's better to try a few lines from your talk. If you can't get into the room before the audience arrives, find out as much about it as you can.

If all this preparation fails to prevent the physiological symptoms of nervousness, remember that the audience won't be able to see that you are shaking, unless you are clinging to the lectern and are a victim of resonance. Your nervousness is always more apparent to you than it is to your audience, and what the audience can't see shouldn't worry you.

Speaking out

George Bernard Shaw once described a nap as 'a brief period of sleep which overtakes superannuated persons when they endeavour to entertain unwelcome visitors or to listen to scientific lectures.' Sleep is the harshest criticism a scientist on stage is likely to face. This doesn't mean that the science dulls the senses: the audience may fall asleep not because your ideas are boring or difficult, but because you are inaudible. Before people can listen they have to be able to hear.

Some people assume that the way to be heard in a large room is to shout, but all this actually achieves is a shell-shocked audience and a sore throat. To get your message across with the minimum damage and maximum effect, you need to master some of the skills that actors use to fill huge theatres night after night, without ruining their voices: breathing, voice production and articulation.

Breathing

It is difficult to think about breathing – it is, after all, something we do all the time without thinking. However, in the same way as you can control your breathing for some special activity like swimming or singing, you can also control it for public speaking, and use it to produce an expressive and interesting voice. To do this, you need to know something about the anatomy of breathing.

The intercostal muscles lie between the ribs. When they contract they pull the ribs upwards and outwards, making the chest expand. This decreases the air pressure in the lungs, and you automatically inhale. As the intercostal muscles relax, the ribs fall back and push air out of the lungs. This is the air you use for speaking. The diaphragm is a dome-shaped muscle which lies above the stomach and just below the lungs. When it is relaxed it bulges upwards, but when it contracts it flattens slightly, so that the chest has more room to expand. One way to gauge the importance of the diaphragm in breathing is to think about the effects of a big meal. Eating too much before you speak can sabotage your careful preparations on several fronts, one of which is that a full stomach leaves the diaphragm very little room to manoeuvre, and so reduces the space available for the chest to expand. You quickly run out of breath.

The diaphragm and the intercostal muscles act together involuntarily to keep us breathing. There are, however, other ways of breathing which use voluntary muscles, and so are easier to control. The muscles which lie across the abdomen just below the stomach – they are sometimes called the stomach muscles – can be used to put pressure on the diaphragm. By pulling in with your stomach muscles you can push the diaphragm upwards, which gives you extra breath to exhale. By relaxing them you allow the diaphragm more room to flatten out, thereby making more room for the chest to expand.

Another way of getting air into the lungs is to raise the shoulders. This method is very inefficient, tiring, and can damage your voice. By raising the shoulders you squeeze the chest into a long and narrow shape so that the diaphragm can't do its job properly, and only a small amount of air reaches the lungs. You also put pressure on your throat so that speaking is difficult, and you will look and feel very tense. If you find your shoulders creeping towards your ears, wriggle your shoulders until they drop into a more relaxed position.

Knowing where all these muscles are is not enough: you have to learn to control them. By practising slow deep breathing you can feel where the muscles are and what they are doing, and then they are easier to control. Try lying on your back, completely at ease and breathing normally. Keeping your shoulders relaxed, put your hands gently on your ribs and abdomen and feel how they are moving. You will feel your chest expanding and contracting and your stomach muscles moving in and out. Now breathe in very slowly, through your nose, for as long as you can – and then hold the breath for as long as you can. When you can't wait another second breathe out – but slowly – through your mouth, and without making a sound. When your ribs are completely relaxed and you think you've run out of breath, pull in your stomach muscles and you'll find that you've still breath left. When all your breath is gone, breathe in through your nose again. You'll feel your ribs spring out and your stomach muscles relax.

Now go through the process again, counting in your head through each stage. On your first attempt you may be able to count to ten as you breathe in, eight while you hold your breath and five as you breathe out. Try once more, taking longer for each stage, and try to breathe out at the same rate as you breathe in. By repeating this exercise you can learn to use your full capacity and to control your breathing out, so that when you are speaking you always have enough breath when you need it.

This breathing exercise is very relaxing. You may even find that you fall asleep when you try it. You can use it to help you relax before you speak – even sitting in your chair on the platform you can take three or four slow deep breaths while your host is introducing you.

Voice production

To produce a voice you turn your breath into sound. You make sound easily enough, by passing air through the larynx. This sound is amplified and shaped by resonance, which occurs in your chest, mouth, throat and nose. The more resonance you can manage, the stronger your voice will be. As with musical instruments, the more space you have for resonance the louder and stronger the sound you will produce. However, whether you are one of nature's tubas or a tin whistle, there are always ways of finding extra space for resonance.

Breathing deeply helps – your voice comes in part from the spaces in your lungs. But the most important resonance for speaking occurs in the mouth and throat. This is why it is so important to open your mouth adequately when you speak – an open mouth provides a bigger space for resonance. It is also one of the reasons why you should look up and at your audience, for if you address your feet you will not only lose a lot of volume into the floorboards, but you will also constrict your throat, giving less space for resonance. Opening your mouth means moving your jaw. Watch yourself speaking in a mirror and see how much your jaw moves. Now try exaggerating the movement of your jaw and you'll hear the difference. If moving your jaw like this leaves you with an aching face, then you have muscles which can help you when you speak which are in need of exercise. One way to exercise these muscles is to repeat the word 'bananas'. Say 'baah-naah-naahs', keeping your face relaxed and just dropping your jaw to open your mouth as far as you can on the long aahs. Do this for a few minutes every day – you'll look ridiculous but you'll feel and sound much better. Another way to loosen your jaw is by yawning – but if you choose this method make sure you practise before taking the platform.

The nose and sinuses are useful resonators, as you'll notice when you have a cold. However, on their own they give the voice a nasal tone which doesn't carry well or sound good. To avoid a nasal tone, open your mouth properly when you speak. That way you will be resonating your voice in your mouth and throat as well as in your nose, which will sound much better. You can tell if you have achieved this by touching your face and neck – you can feel the vibrations.

Resonance is well worth thinking about and practising; it can make an ordinary voice sound great, as anyone who sings in the bathroom will know. Singing is actually a good way of improving your resonance, especially if you can find somewhere where you can cast aside your inhibitions and belt out your favourite tunes without worrying about who might be listening. The bath is an ideal place for singing, since the humidity is kinder to your throat than dry air. Try this: chose an easy tune and pretend you are an orchestra. Sing your tune as though you are a flute and your voice will resound in your nose. Pretend you are an oboe and the sound will come out of your neck. The bassoon will be somewhere in your chest. Once you've got through the entire orchestra you'll have found all your resonance spaces, and once you know where they are you'll find that you use them.

Singing is also a good breathing exercise, and is relaxing. You could sing along to your favourite tape on your way to the talk – though perhaps not, if you plan to use public transport . . .

Articulation

Articulation is the process of turning sound into words. The vowels will look after themselves, providing your resonance is adequate and your jaw well exercised. Making vowels is like saying 'aaah' for the doctor – you just open your mouth and make a sound. Consonants require more effort.

Tape yourself speaking and listen carefully to your consonants. If they aren't clear, repeat them until you determine which parts of your mouth you use to make the sound, and then think about exactly what you are doing as you repeat the sound. This will tell you which part of your mouth you need to exercise in order to make your consonants clearer.

Articulation uses almost every organ of your mouth and face. The most important, though, are the tongue and lips. Tongues are difficult to exercise, but you could try this. Push your tongue out as far as you can, and then try to touch your nose and chin with the tongue tip. Breathe through your nose while you do this. A rather more polite exercise involves making the noises 't', 'd', 'l' in that order. You should find when you do this that your tongue is behind your top teeth for 't', on the ridge behind your teeth for 'd' and on the roof of your mouth for 'l'. Once you know these positions you can do this exercise without making the sounds, by just moving your tongue.

Lips require a little more singing. Try singing 'many men' over and over again very quickly to a fast moving tune – the William Tell Overture works well. It goes like this: many men, many men, many men men men, many men, many men, many men men men, many men, many men, many men men men, many me-e-e-e-en, many men men men. . . .

Giving talks with a untrained or unpractised voice is like running a marathon without training. Your voice is produced by a very delicate part of your body, so don't do anything with it that hurts. Happily, most people don't start their speaking careers by making a thirty-date lecture tour, but if your research post suddenly turns into a lectureship or you get a sore throat when you speak, see a voice teacher. He or she will be able to find a set of exercises that will suit you, and will help you to find a way of keeping your voice in trim. Your public library or nightschool should be able to put you in touch with a teacher, but make sure he or she is qualified to teach – a qualified teacher will be affiliated to or licensed by a drama academy or professional society.

Gesture and posture

One popular guide to 'effective speaking' contains a full chapter on gesture with long descriptions of where to place your arms, how to curl your toes, bend your fingers, raise your eyebrows and stand, sit and walk. This approach recalls the books on elocution and oratory which were popular a hundred years ago, which gave formulas for 'rage', 'passion', 'hypocrisy' and 'sincerity', and provided a repertoire of gestures: a pointing finger meant a warning and a clenched fist a threat.

Nowadays we are inclined to leave such stylized gestures to traffic policemen. The comparative intimacy provided by microphones and well designed theatres means that we can afford to be a little more subtle. Think of gestures as an extension of your voice and of your vocabulary. The good gesture accompanies what is being said, illustrates it and emphasizes it. Try a gesture appropriate for 'the vibrations get out of step' or 'the cell envelopes the food particle'. Too many gestures can be a distraction and lose their impact: the third time you thump the table the audience will assume you are crying wolf and may ignore your most important point. After all, gestures should be meaningful, to both you and your listeners. If they don't fit happily with your words, they will confuse the audience. As the Victorian actor Henry Irving said, 'It is easier to detect a flaw in an actor's impersonation than an improbability in a book . . . a false intonation is immediately on the ear, an unnatural look or gesture is promptly convicted by the eye.'

Bearing this in mind, the question 'what shall I do with my hands?' is a psychological rather than a physiological one. If you concentrate on what you are saying, you'll not notice what you are doing with your hands, and what's more important, neither will your listeners. During the course of your talk one hand may find refuge in your pocket, which is fine if this is where it wants to be.

There is no 'correct' posture, although some books think there is. One commands:

> Stand upright! Keep chin level and shoulders level. One foot, either the left or the right, slightly in advance of the other. Knees not to be bent. Heels slightly apart. Weight of the body to be on the toes, rather more on the advanced than on the rear foot. Feet open at an angle of 45 degrees.

This is not easy to achieve, particularly if you have forgotten your protractor. But there is a good posture, and it combines – in the military jargon – both the 'at ease' and the 'attention' commands. You will then be able to suggest to the audience a combination of poise and alertness. The audience will feel comfortable if you look as though you are, so find a posture which suits you. It is important not to slouch, since this will impede your voice and make you look tired, and try not to rock from side to side. This can put an audience to sleep – it certainly works with babies.

If you are using a microphone make sure it is at the right height and in the right position for you. Try not to lean on it, drape yourself over it, or sway to and fro like a fire-and-brimstone preacher. If there is a single stem microphone in the middle of the stage you will have to walk up to it. If you are nervous you will feel like Marie Antoinette heading for the guillotine, but don't rush – a slow, unostentatious walk is all that's needed. If it is fixed to you test it to check that it is in the right position – you may find that the only noise it amplifies is your heartbeat. If there's electric cable involved, check that you have enough of it, or you may find that your casual stroll to the blackboard is brought to a premature and very sudden halt.

Deliveries

Being clear, audible and at ease is not enough to make a successful presentation. You have to sound interesting as well, and from the very moment you begin. Show your enthusiasm for your subject – enthusiasm really is infectious, and you can't hope to sweep people off their feet if you won't be swept off your own.

Looking at your audience is vital. It holds their attention, and helps them to feel part of the event. Look at everyone – the people at the front and the back of the hall and those in the middle and at the sides. If you look out of the window, or at the ceiling, so will your audience. You can encourage your audience to look at a slide or model by looking at it yourself, but this need take only a couple of seconds: after that the audience will want to look at you as well as the picture, so you must look at them. If you don't, the audience can quickly forget that you are there, and their attention will soon wander. By looking at the audience you can judge their mood. Are they enraptured, or should you try to be a little more enthusiastic? Are they looking sceptical and in need of persuasion, or are they agreeing with you already? Be careful not to cling too tightly to any glimmer of response: if some poor soul makes the mistake of nodding in agreement with a point of yours, you should not punish him with a look that pins him to his seat – the people who aren't responding need your attention too. Try to move your head around naturally, as you might in a discussion around a table. People appreciate being looked at, but grow uncomfortable if the glance turns into a stare.

An even delivery can distance your listeners, by boring them if it is understated and by exhausting them if it is over-enthusiastic. We have all experienced music which has sent us to sleep because it is repetitive or monotonous, and have been startled to attention by an unexpected flourish from the brass section. The same is true of voices. If you listen to a good speaker – a newsreader or an actor – you will hear that the voice is continuously changing. It was said of Gladstone that 'if you studied the use of his voice you would see that it was like a great gamut of music, where the soft note and the low note came just at the right moment. To listen to him was like listening to a wonderful piece of varied music'. Whilst no-one expects you to produce a symphony in speech, a little well-placed musicality can be very effective.

You will probably be surprised at the sound of your own voice if you record yourself speaking. Listen for any variations of pitch and pace, and then record yourself again, this time exaggerating the variations. Try saying a simple sentence like, 'I called, but there was no reply' as though you were relieved, then sad, confident, happy, nervous, indignant, angry, pleased, tired, and panic-stricken. Listen to the difference in your voice each time. This exercise can give you an idea of the variety you can achieve, and will show you how to use pitch and pace to express different feelings.

There is plenty of scope for putting variety into your voice when talking about science. When you started your investigation you were unsure; when your equipment broke you were cross; when your results weren't what you'd hoped for you were disappointed; when you reached a satisfactory conclusion you were overjoyed. Remember how you felt when your paper was accepted? When the cleaner threw away your experiment? When the laboratory roof collapsed? When funding for your work was increased? Share these feelings with your audience – they are an important aspect of your life as a scientist, and by communicating them you will make your voice, your talk and your science more interesting.

Bad habits

An audience is likely to be put off the track by a number of common, but easily cured, bad habits. Speakers are often unaware of their idiosyncrasies, so ask your friends to tell you about yours. Once you know that you tug your earlobe or hum between paragraphs you probably won't do so any more.

There are three common bad habits which are certain to distract your audience, irritate them or lose them completely. The first is the 'um' and 'er' habit, with its many associated brethren. Rarely is even a good talk completely free of the 'ums' and 'ers' of everyday indecision; for a few moments, supply of intelligent material from the brain is interrupted. In such circumstances, many speakers feel they have to keep up the flow of words, as if to assure the audience that their vocal apparatus is still active. Actually, it is better for you and for your listeners if you pause, giving them time to catch up and you time to collect your thoughts.

It is not easy to abandon the comforting 'um' for a lonely silence, but this exercise may help. Look around you and describe what you see for about a minute, taking care not to 'um' or 'er', but pausing where appropriate. If you let an 'um' slip out, start again, and do the exercise as many times as it takes. Get a friend – a patient friend – to listen to you. You could change the exercise slightly: find someone to interview you about the news, or describe a breakthrough in your field. The aim remains the same: to train yourself not to waste words and energy.

Faraday's laws of public speaking

1. Never repeat a phrase.
2. Never go back to correct yourself.
3. If at a loss for a word, simply wait and it will come.
4. Never doubt a correction from the audience.

Michael Faraday

The second habit also wastes energy, but is easier to break. Some people seem to feel it necessary to apologize or make excuses for what is otherwise a perfectly adequate performance, to the irritation of their audience. You can apologize for unavoidable events – your train was late or the slide projector exploded – but never for things you could have put right yourself. If you knew that your overhead transparencies were illegible you should have made new ones. If you knew you couldn't be heard in such a large room, or if you didn't have time to prepare properly, you shouldn't have agreed to give the talk.

There are, lastly, the physical equivalents of the 'er' habit: fidgets, wriggles and other mannerisms which can distract your audience. Some mannerisms can be endearing, and from Aristotle's custom of walking about as he lectured to mathematician Felix Klein's much imitated practice of spearing his cigar on the end of a penknife and waving it aloft, scientists have cultivated them for effect. But be warned: many people find such habits irritating. If it is your pleasure to twiddle keys or coins or pencils, or to perform any other conjuring tricks with inanimate objects, you must force yourself to stop even if it means not taking them with you to the platform. If you concentrate hard on what you are trying to say, you'll find your spare energy goes into a vigorous performance, where it belongs.

These habits are worth thinking about because they place barriers between you and your audience, making communication more difficult than it need be. The preparation which you have put into your talk is designed to convey ideas and information in the right sequence and with the appropriate amount of detail. The delivery, if it is effective, will ensure that the ideas and information arrive at their destination. Anything that impedes that journey needs attention.

Coping with questions

Questions usually follow a talk, but they can accompany it or even interrupt it. You may, with the chairman's prior agreement, offer to take questions as you speak, rather than at the end. This has the advantage that you will keep your audience hooked even if you have to deal with difficult questions; the disadvantage is that you may lose track, or find yourself diverted from the main theme of your talk by a barrage of irrelevant queries. If you are not supremely confident it is probably better to leave questions from the audience until the end.

You can get your listeners to sit up and take notice during your talk by asking them questions. 'What on earth could we possibly want with yet another 47 000 high velocity protons?' 'So the plant was dead, there were six inches of water on the lab floor, there was a power cut and it was 7.30 pm on Christmas Eve. What do you think we did?' Don't be surprised if someone shouts out an answer – it will show that they were listening and will lighten the atmosphere.

Questions are usually invited at the end of a talk. If this invitation is met by silence, a kindly chairman will probably ask a question. Wherever it originates, a question is not a licence to give another speech, so your answer should be brief. If you don't know the answer, say so, but you can offer to find out. If someone is asking a stream of questions, offer to discuss the matter later and move on to someone else. You may find that you are asked a triple-decker: three questions in one breath. 'I'd like to know first . . . also . . . and thirdly . . . ?' These are tough: you have to listen, make notes, and prepare some kind of answer while you are listening to the question. See if you can fit the questions together and give a general answer. Otherwise you'll just have to answer them swiftly and in order – which is why it is useful to note down the questions.

When questions arise, respond to them with tact. A question is more likely to be a polite enquiry than a criticism, but speakers sometimes hear more hostility in a question than is meant – your questioner may be nervous and use words which sound sharper than intended. 'What about the points on the graph above and below the line you drew?' does not necessarily imply that you cooked the books. You may be able to predict some of the questions, so discuss them with colleagues before you give your talk. A foolish question needs to be dealt with gently – your listeners will come to their own conclusion about its worth.

Encore!

One of the best ways to improve as a public speaker is to watch others, looking for their accomplishments as well as their faults. Standards in science aren't always high, so look elsewhere – in politics, business, the church, radio and television. The French naturalist Georges Cuvier learnt his public speaking by going to the theatre and watching the most accomplished actors of his day, and was soon attracting enthusiastic audiences.

More recently, an opera singer made his debut at La Scala. After his first aria the crowd called out 'Again, again!' Delighted by the reception, he sang the aria a second time, and again they called for an encore. The exercise was repeated a third time and a fourth . . . Finally, panting and exhausted, he asked, 'How many times must I sing this aria?' The audience's reply was 'Until you get it right!' That's how it is with public speaking – but do your practice in private, and, as they say in theatrical circles, it will be all right on the night.

Recommended reading

Stephen D. Lucas, 1989, *The Art of Public Speaking,* third edition (New York: Random House). This is written for college students enrolled in public speaking classes, but contains a wealth of information for all speakers on such subjects as preparation, presentation, visual aids and selecting topics.

Gordon Luck, 1982, *A Guide to Practical Speech Training* (London: Barrie & Jenkins). Skip the theory and use the exercises.

Michael McCallion, 1988, *The Voice Book* (London: Faber). Written for actors, but the index will guide you to the comprehensive coverage of the basics.

References

1 Michael Elder *et al.*, 1989, *The Oxford Textbook of Psychiatry*, second edition (Oxford: Oxford University Press).
2 Richard P. Feynman, 1989, *Surely You're Joking Mr Feynman!* (London: Unwin Hyman).

3.3 Visual Aids

Technology has been a great help to the public speaker since the turn of the century, when the magic lantern was first invented. The magic lantern was a light-proof black box containing a lamp and a lens which projected an enlarged image onto a screen. People went wild with excitement and queued for hours to see it, and ever since then we have been amazed and delighted by pictures in light. The magic lantern has had its day, but there is now a huge armoury of visual aids which are far more versatile, and which provide a wide range of methods of communicating science.

Show and tell

We learn and remember far more by looking than by listening. Several years ago the US Air Force tested people's ability to see, hear, absorb and retain information communicated to them by words only, and by words and pictures. It was found that seventy-two hours after a presentation with words and pictures, an audience remembered many times more than when only words were used. When these results were publicized, a wave of enthusiasm for visual aids of all sorts flooded across the campuses and conference halls. For a while it seemed as if pictures alone could take the place of words.

Nowadays a little sanity has been restored, and it is possible to see the potential, as well as some of the pitfalls, of using visual aids in scientific talks. Visual aids can be used for communicating information, as well as providing variety and stimulating interest. The happy conjunction of word and image can be quite spectacular, but pictures can also be distracting and intrusive, particularly if they are not quite right. Fit your visual aids to your talk rather than the other way around, however stunning your slides or model might be. The first thing to ask about visual aids is whether they are really necessary – they should not be used as props to hold up a poor presentation, nor to force a point. It is better to attend first to the words, and then decide where to use visual aids. Pictures cannot give your talk for you. As biologist Peter Medawar put it: 'If the audience won't take a scientist's word for it, they won't take his slide for it, either.'

Visual aids can be signposts in a talk, the equivalent of headings and sub-headings, to signal a change of pace or subject matter. A slide or a model can add emphasis (the equivalent of saying 'I think it is important to remember . . .') or act as a break point (like saying 'to look at the matter in another way . . . ').

Some advice applies to any visual aids you might use. They should be as simple as possible. Put the bare minimum on diagrams and elaborate out loud. If you are showing a table of figures point out trends – higher conductivity at lower concentrations, the peak in 1975. A graph can do this for you, and will have a more immediate impact than a lot of numbers. If you show a graph, explain the axes and units, and ensure that any lettering is legible from the appropriate distance. The audience will spend time reading a picture, and a good way to judge how long this will take is to read it yourself. Pictures should not contain more information than your audience can be expected to assimilate in a very short time, with or without your own spoken 'captions'. The audience is seeing the picture for the first time and listening to you as well: that's a lot of data reaching the brain at once via ear and eye. If it takes you more than a couple of minutes to read a picture, you have put too much detail into it. On the other hand, if you can take in a picture at a glance, it may be superfluous – ask yourself whether the information couldn't be conveyed more easily by speech. You will have heard of the speaker who presents a slide show which begins with a startling black and white GOOD AFTERNOON and then proceeds, after a few words of introduction, to the main points (slide: MAIN POINTS) of which there are five (next slide: FIVE POINTS), which are each treated individually (clink, clunk: POINT ONE), until he gets to the end – or rather THE END. This speaker exists and is regularly sighted: please do not encourage him.

Making good pictures is partly a matter of ensuring good quality reproduction. You may find that your photographic department can advise on how much detail to include, and on how clear a diagram or picture will be on screen. It is not always possible to take a picture from one medium and present it in another. Diagrams from books don't usually make good slides – in the book a diagram can be referred to time and time again at the reader's leisure, and so can contain a lot of information. To make an effective slide you will have to reproduce the picture in a simplified form, highlighting the important parts and deleting irrelevant details.

When you first display a visual aid, look at it to check that the right picture is being shown, and that it is level, the right way round and in the centre of the screen. This is particularly important with overhead transparencies, which get wafted askew by the wave of a hand. Then look at the audience. If they are screwing up their eyes or looking baffled, ask them what's wrong. You may be out of focus. All this looking will give your audience a valuable few seconds to study the picture before you start talking about it. You should be able to talk about the picture without looking at it, so that you can face the audience when you speak.

Four don'ts: don't use visual aids simply for scenery, or the audience will wonder what they are and why they are there. Don't display a picture or model before you are ready to talk about it, or the audience will look at it and not listen to you; and don't leave it on show after you've finished talking about it for the same reason. Don't use visual aids in quick succession without a few words in between, or the audience will never get to know you, the speaker.

If you are nervous, visual aids can be a great asset. Whatever it shows, if you put up a slide or an overhead transparency the audience will look at it, rather than at you. If you do feel uneasy during the first few minutes of your talk, put up a picture quite early on, and the audience can look at that while you collect yourself. If you are worried that your notes won't be enough to keep you on the right track, use the pictures. Put labels 1, 2, 3, . . . on a diagram and describe the parts in that order, and you won't miss anything out. Write a list of points on an overhead transparency and reveal them one by one, explaining each as you go along.

Find out from the organizers before you plan your talk what equipment will be available. Take anything portable – spare transparencies, pens, a pointer – with you, just in case.

Slides

'Could I have the first slide please Mr Johnson – Oh! I think you must have the wrong box. No, no – that's right! – it's just that the slides are in the wrong order. If you could just turn them back to front, I mean, er, put the last one first and the first one last. Oh dear . . . Well, anyway the first slide wasn't important anyway. Ah! There it is. As you can see, it's, um, er, yes it's upside-down. Anyhow, you can get the gist of it. You see the diagram shows the relationship between particle size and energy; and there's a cluster there on the top right, or rather – the top left . . .'

Sounds familiar? Slide projectors – any visual aids in fact – are more complicated than many people realize. They have spectacular potential if they are used properly, and several inherent limitations which make it difficult to get the best effect from slides and easy to get the worst. The slide projector is not an iron lung designed to resuscitate a talk that is teetering on the verge of collapse, though many speakers look upon it as a source of artificial respiration, brought in at a moment of crisis. The good news is that slides can add colour and excitement, and at the same time they can help to express difficult ideas and data in an easily digestible form. When spectators at a baseball match cried out 'Slide, Kelly, slide', they weren't calling for another picture but for a piece of acrobatics that could make the different between a run and an out. Visual slides can make a similar difference.

If the projector is not fixed in place or has not been set up before you arrive, spend a couple of minutes checking that it's in a good position. Carry with you a 'test slide' which you can use to focus the image and check the size of the picture even if your listeners are already settling down in their seats. What goes on to your test slide is up to you – it might be an all-purpose image with 'TEST SLIDE' written against a colourful background. Don't use a slide from your talk, or it won't be a surprise when you use it for real.

There are two ways of changing slides. The better way is to use a hand-held control and work the projector yourself. If you do this, practise before you talk. The other is to delegate the job, in which case you should try to avoid the 'next slide please Mr Johnson' routine by planning in advance. If you give the projectionist a copy of your notes with the visual aid instructions clearly marked, then during your talk a well-timed nod in the right direction will produce the appropriate slide.

Finally, your slides should be numbered in the top right-hand corner and numbered in your notes. During the question and answer period you may want to call up a slide; it's far easier to do so by asking Mr Johnson to take out slide 8 than by running backwards through the whole set.

Overhead projectors

Overhead projectors were originally intended to be used by academics who stood at floor level with magnified pictures projected high above their heads in large lecture theatres. The speaker could face his audience, point out particular features of an illustration and sketch in new data as the lecture progressed. What happens now is that rooms are too low to project the image properly, the projector often stands on a table and it is difficult to avoid standing in the way. If you look up at the screen once in a while, you will able to see if the audience is looking at your magnified silhouette.

Lettering should be simple, clear and legible, and if your handwriting isn't, prepare transparencies in advance using a stencil. As a rough guide, letters and symbols need to be about one inch high on your transparency to be read at a distance of about thirty feet when projected.

Colour in transparencies adds greatly to their impact, but you will need a much darker room and a brighter screen than for black-and-white. Red, yellow and orange don't show up well, but most other colours are clear enough. One of the advantages of overhead projectors is that you can point out details on the slide and the image of your pointer will appear on the screen – magnified. One disadvantage is that if you are nervous and pointing with a stick or pencil, your shaking will be magnified first by the pencil and then by the projector. To avoid this you can touch the slide with the pointer and hope that this keeps it still. If that's not enough, point with just a finger – but try not to touch the slide, as your finger may smudge your drawing.

When you have finished with the projector, or if you have a long gap between slides, switch it off. People will stare at a lit screen even if there is nothing on it, and if there is nothing on it they ought to be looking at you. Beware though – some projectors take ages to warm up, and do so noisily. Try yours before you start. If your projector doesn't respond quickly, leave it on throughout your talk, but block out the light when you don't need it by putting a piece of paper where the slide should be.

Flip charts

If you plan to use a flip chart you will have to provide it yourself, but check that you'll have somewhere to hang it when you speak. Flip charts can be useful if the room is small enough for everyone to read them, and if you write legibly and boldly. Eastman Kodak's *Legibility Standards for Projected Material* advises: 'To ensure accurate recognition of symbols they should subtend at least 9 minute of arc from the position of the farthest viewer.' In plain English that means that letters and symbols will be visible if their height is 1/32 inch for every foot away. A one inch letter is legible from 32 feet; a ½ inch letter from 16 feet and so on. That's the theory. In practice, and since not everyone has 20-20 vision, your letters should be at least two inches high to be visible at 30 feet. If the room is 60 feet long, make them 4 inches high. If it's 100 feet long, don't use a flip chart.

Write on your charts before you start, and leave every other page blank. That way you can get rid of pages you've finished with and still only reveal a new page when you are ready. Or you could cheat a little: draw lightly in pencil on the chart beforehand, and then draw over the pencil marks with a marker pen during your presentation.

While colour is useful, bear in mind that some colours – yellow, for example – do not stand out well, and that others clash. A striking design will be more powerful than a riot of colour: tabloid newspapers with their bold headlines, boxes and contrasts offer an excellent model.

Blackboards

'Chalk and talk' has a bad name. Blackboards are messy, and remind people of school. But they do have their uses – people who are nervous or find it difficult to keep still are relaxed by the physical exercise of writing on a blackboard. If you want to improvise – say you have asked your audience for ten reasons why they didn't take science at school and you want to list them – the blackboard is ideal, as you can write on it quickly, correct your mistakes and wipe it clean when you've finished.

The disadvantages of blackboards are that it is difficult to write on them legibly, and that while you write you have your back to the audience. However, if your handwriting is clear enough, and if you can keep in touch with the audience while you write, you may be able to put a blackboard to good use.

Films

A film can be a high point in any presentation, but keep it short. The audience has come to see you, not a movie. Timing is all important: don't show a piece of film at the beginning of your talk, or your audience may not appreciate its significance. More importantly, what follows – you – may seem something of an anticlimax. However ebullient and dynamic you are, you'll have a hard task competing with a well-made film.

You can achieve some spectacular special effects by not running through the film and checking the projector first. There is the 'stop and start' effect produced by a broken piece of film: one minute a picture, the next a blank screen. There is the 'sudden exploding projector bulb' effect, which plunges the hall into darkness and leaves only the muttered expletives of the projectionist to fill the void. Always run through the film, check the spools, ensure that a spare projector lamp and splicing equipment are available, and make sure that someone knows how to dim the lights, start the film, stop the film and raise the lights – in that order.

Video

Video cassettes are easier to use than 16mm films, but the picture size and quality is rarely as good. Different institutions use different equipment and a VHS cassette cannot be shown on a recorder designed for Betamax film, and vice versa. Different countries use different colour systems: an American VHS cassette will not work in British or Australian VHS equipment, or in any European equipment. However, many large institutions will have the necessary recording and projection equipment, though you should always specify your requirements well beforehand. A normal sized TV monitor can be seen at a maximum distance of 40 feet, so if that's not enough, see whether equipment is available for projecting the video image onto a large screen. Though this takes some time to set up, it is well worth the effort.

If you have more than one piece of video film to show, transfer all the clips onto one cassette in the correct order and with clear and sharp breaks between sections. The alternative is either to use several cassettes, which is awkward, or to rely on the counter on the recorder to guide you through a single cassette, which is risky since recorders use different counting systems. Keeps the clips short: if the audience wanted to spend the evening watching television they would have stayed at home.

Demonstrations

Scientists are used to seeing extraordinary things happen. Living cells divide, sturdy metals dissolve, and invisible particles form visible vapour trails through apparently empty space. Most people never witness such things. Experiment is something which happens behind closed laboratory doors, and only theory occasionally creeps out. Yet many people first discover the excitement of science when they see something strange, and then ask the all-important question 'how did that happen?'

Throughout history, simple and elegant demonstrations have been used to great effect. Archimedes pleased Heiron when he demonstrated his giant catapults; more recently, Richard Feynman embarrassed a committee investigating the causes of the 1986 NASA shuttle explosion by demonstrating that a rubber component stiffened at low temperatures. He dropped the component into a glass of iced water, and retrieved it, for all to see, in the middle of a press conference.

Two centuries ago, public interest in the sciences was awakened by demonstrations, and experiments were performed for paying audiences. A newspaper advertisement of 1746 invited readers to attend a series of 'Experiments with Concise Observations by Way of Lecture' at London's Swan Tavern, where 'the whole Affair of Electricity will be pursued from its Simplest to its most capital Phenomena'.

Later the Royal Institution established a lecture theatre where ordinary people could hear about science and witness a wide range of demonstrations. Some of its lecturers became known to a wider public for their demonstrations as much as for their scientific achievements. Thomas Young demonstrated wave interference there; Humphry Davy focused heat radiation, Michael Faraday made a bubble of hydrogen and oxygen from electrolysed water and exploded it, and James Dewar picked up liquid oxygen with an electromagnet. Now the Royal Institution lectures can be seen on British television, and attract large audiences.

Yet only a small minority of scientists use demonstrations in their public talks. The reason for this is that speakers worry about whether their demonstration will 'work', but there are countless simple experiments which never fail and which entertain and enlighten an audience. The experiments you remember from school – pondweed producing oxygen, Brownian motion in smoke, solutions changing colour with pH – may seem commonplace because they are straightforward, and yet they are of great interest to someone seeing them for the first time.

Seen and not heard

> It is surprising how often people in all walks of life own that their interest in science was first aroused by attending one of these courses (at the Royal Institution) when they were young, and in recalling their impressions they almost invariably say not 'we were told' but 'we were *shown*' this or that.
>
> Sir Lawrence Bragg, physicist

Handling the goods has become an increasingly popular method of instruction in 'hands on' galleries in science museums throughout the world. Visitors do not gaze silently and read labels; they pull levers, press buttons, assemble and destroy. This is vandalism in a good cause, combining education and entertainment. Providing they are big enough, a hands-on gadget can be a useful visual aid. You can involve the audience by asking for volunteers to work the apparatus. You can get ideas for demonstrations by visiting science museums which have 'exploratories' or hands-on galleries. Such places usually have shops where you can buy replicas of the exhibits and scientific toys to use in your talk. If you don't have a museum to hand, try your local toyshop.

While these devices can make a talk more exciting or easier to understand, they are never the only way of getting your message across. If your talk is a slick succession of high-tech images your audience will think you have sent along your laboratory, and won't notice you in the midst of all the clutter and lights. The star of the show should always be you, and not your slide projector.

Recommended reading

David Barlex and Clive Carre, 1985, *Visual Communication in Science* (Cambridge University Press). Written for teachers, it shows how visial aids can be used to make science clearer and more fun.

Richard Gregory, 1986, *Hands-on Science: An Introduction to the Bristol Exploratory* (London: Duckworth). Written by a pioneer of interactive exhibits, the book shows how scientific ideas can be demonstrated, as well as explained.

Charles Taylor, 1988, *The Art and Science of Lecture Demonstration* (Bristol: Adam Hilger). Full of practical ideas for demonstrations.

IV. *M*eeting 'the media'

Meeting 'the media'

In May 1911 Rutherford published a paper in *Philosophical Magazine* entitled 'The Scattering of α and β Particles by Matter and the Structure of the Atom'. It announced what we now call 'the Rutherford atom': a tiny, heavy, positive nucleus with negative electrons in distant orbit. But the *Philosophical Magazine* paper was not our first glimpse of the Rutherford atom: Rutherford himself had given a talk on it two months earlier at the Manchester Literary and Philosophical Society. The talk was reported in the *Manchester Guardian*, but the journalist had left the talk believing that the positive and negative regions of the atom were of the same order of magnitude, and the revolution passed unnoticed.

In 1971 the *New York Times*' science correspondent Walter Sullivan reported some work that had been done jointly at the Stanford Linear Accelerator Center by researchers from Stanford and the Massachusetts Institute of Technology. When the article appeared, only one MIT scientist was mentioned – Victor Weisskopf – and only as a commentator. The MIT scientists were upset, but were unwilling to believe that their Stanford colleagues had deliberately left them out of the picture. What had happened was this: Weisskopf had telephoned Sullivan, and had told him that there might be a story for him at Stanford. Sullivan rang the team at Stanford, and said that MIT had suggested he call. The scientists at Stanford assumed that Sullivan had already got the MIT details, and gave him their side of the story. Since Sullivan never knew about the MIT contribution, he wasn't able to report it. A letter setting the record straight was published a few days later, but by that time the excitement had passed, and the authoritative *New York Times* had already given Stanford the PR boost that it should have shared with MIT. At least, that's the story according to Michael Riordan, an MIT scientist who was working at Stanford at the time.[1] Other accounts give a different picture: Wolfgang Panofsky, director of the Stanford accelerator, is reported as saying that Sullivan was given and did include all the names of the MIT contributors, but his editor cut them out.[2]

Despite the good intentions of the reporter who listened to Rutherford's talk, and of the scientists who spoke to Walter Sullivan – the USA's most respected science journalist – the stories didn't turn out well. Even Sullivan's experience and care didn't protect him from incomplete information or – depending on which account you read – a ruthless editor. On other occasions experience and care are in short supply, and science stories are the product of a hazardous mix: a scientist who is ignorant of the ways of the media and a journalist who neither knows nor cares about science. There are now an increasing number of science graduates in science journalism, but there are still some scientists who need to learn about the media.

Scientists and science correspondents have long reported for newspapers, providing both news and features. Attitudes change: in the 1830s *The Times* mocked the new British Association for the Advancement of Science, referring to it as the 'British Ass', and refused to carry any reports or announcements of the meetings except as paid advertisements. Now the British and American Association meetings are awash with journalists, and, for a few days at least, science is a major feature in the newspapers. Science took to the airwaves in 1942 on the BBC's Home Service radio, when housewives struggling to bring up baby in the austerity of wartime Britain were given advice about nutrition and health by the Radio Doctor, who, with his bluff manner and clear expression in everyday language, was welcomed into the homes of millions. Science programmes appeared surprisingly early in the relatively short history of television: the first was Inventor's Club, which was broadcast from 1948. The BBC announced 'a promising and important new project . . . to find the latent inventive genius in this country'. A panel of experts commented on inventions submitted by members of the public, and the producer hoped that returning servicemen, full of new experience and a wider view of the world, might boost the flagging post-war economy by putting their creative powers to work for the export drive.

Despite the long association between scientists and journalists, the many differences between the demands of the two professions can still lead to misunderstandings. Scientists see science as a cumulative, co-operative enterprise; journalists like to write about individual scientists who have made a revolutionary breakthrough. Journalists like controversy; science thrives on consensus. Journalists like new, even tentative results with exciting potential; scientists prefer their results to go through the slow process of peer review and settle into a quiet, moderate niche in the scientific literature – by which time journalists are no longer interested. Scientists think that accuracy means giving one authoritative account; journalists feel that differing views add up to a more complete picture. Journalists' work has to fit the space available; scientists' academic papers can be any length. Scientists work at the pace imposed by the nature of the research; journalists are in a hurry to meet a deadline. Scientists must qualify and reference their work; journalists have to get to the point. Scientists want to educate; journalists want to spread news. Scientists are worried about what their colleagues will think, rather than what information the public will receive: scientists' work is judged by their peers, while journalists' work is judged first by their editors

– who may not be interested in science – and then by their readers.

Research costs are large, and budgets are small and administered by politicians. This means that science has become a matter of public policy, and scientists are concerned about the public image of themselves and their work. They are also increasingly aware that this image can be hugely affected by the smallest piece of science journalism. In 1985, the physicist and Nobel laureate Kenneth Wilson persuaded the National Science Foundation to fund a very expensive supercomputer programme. The major weapon in his campaign, he said, was a newspaper article in which a scientist was quoted as saying that without the funding the United States would lose its lead in this technology – a simple statement in a newspaper had a great deal of power. Editors know that when a field of research is adequately funded the scientists don't want journalists around distracting them from their work, but when the same research team is short of funds, they'll be desperate for even a few words of positive coverage.

Most science in the media appears not in science programmes or stories, but as it occurs in social and economic stories. In such cases you are unlikely to be interviewed by a journalist who specializes in science, or who knows any science at all. You may find that a non-specialist reporter will think in terms of how the science will affect people's lives, or be interested in the political or economic repercussions, while a reporter trained in science will think more in terms of the scientific consequences. Even a reporter with a science background may seem to be asking 'stupid' questions: they will know what you are talking about, but will be asking questions so that you provide an explanation which is suitable for their less knowledgeable readers.

No one would deny that science can be complicated, and that ideas often take a lot of explaining. Yet despite the fact that it is difficult to be accurate in a concise, sharp science story, and despite scientists' worries about accuracy, surveys have consistently found that the vast majority of scientists who have been interviewed say that the stories which have been written about them and their work have been accurate. However, they thought that the standard of accuracy in science stories generally was not so good. Scientists also claim that science stories are usually less accurate than other stories. Other research shows that around 50% of scientists complain about inaccuracy in science reporting, but that most of these complaints concern omissions, rather than false information – and omissions are an inevitable consequence of the battle for air-time or column inches. Many scientists criticize the headlines, but these are very rarely written by the reporter who writes the article, and reporters object to misleading headlines just as much as scientists.

The ladies and gentlemen of the press have a poor reputation: horror stories of misunderstanding, misrepresentation and downright lies abound. Tabloid headlines and libel trials catch the eye and persist in the memory.

Yet people still buy newspapers, watch television and listen to the radio, and the reason is that a lot of what journalists write and broadcast is excellent. Most journalism is produced quietly and sensibly, and by skilled, qualified people who are aware of their responsibility to the public; and science journalism occupies a valuable position in newspapers and broadcasting schedules all over the world. Despite the high standard of writing by science correspondents and the reassuring endurance and popularity of broadcast science, many scientists still feel that their subject doesn't get the serious attention or column inches that it deserves. Yet journalists say that if they relied solely on scientists themselves to initiate stories, there would be only one quarter of the science in the media that we see today. There may be a number of conspiracies preventing science news reaching the public, but one that can easily be overcome is of scientists whose prejudice against the mass media and whose reluctance to learn about them can make the job of even the ablest science correspondent very difficult indeed. Bad science journalism is inevitably produced when journalists need a scientist's help and, for whatever reason, don't get it. There are ways of extending a helping hand to journalists and of serving your own interests as well; the most helpful is to acquire the skills of a good interviewee and to make yourself available for interview.

There are a number of circumstances in which you might find yourself being asked for an interview. A journalist may be preparing a feature on a subject on which you have particular expert knowledge and may contact you for information and opinions. What you have to say will be only one ingredient among many in the final story. Of your thirty-minute interview, maybe only a couple of quotes will survive. Perhaps it will all be used as unattributed material. On another occasion you may be approached by a journalist who wants to write an article about your own work, so if it is of current public interest you may well find yourself the subject of media attention. If you are to provide the information a reporter needs in a form in which he or she can use it, you need to know the circumstances in which the reporter works. In this section we look at how reporters find their stories and interview their sources, and at how you can help them to present science so that it is both interesting and accurate. Most of what we've said in the preceding chapters will help you with interviews, and most of what applies to dealing with the press also applies to television and radio.

References

1 Michael Riordan, 1987, *The Hunting of the Quark: a True Story of Modern Physics* (New York: Simon & Schuster/Touchstone).

2 Rae Goodell, 1977, *The Visible Scientists* (Boston, Toronto: Little Brown).

4.1 Interviews for the press

How would you react to this situation? A telephone call comes quite unexpectedly from a reporter who clearly doesn't know the difference between a virus and bacteria, who needs to write five hundred words on your research in immunology by tomorrow. Can he visit you this afternoon so that you can help him get his story straight?

A firm 'no' is easy enough and doubtless the reporter will find another scientist; failing that, he'll struggle for a while and then write another unsatisfactory story about science. You've saved yourself some time, but have done nothing to help the reporter – a professional trying to do his job – or to bring science, accurately and authoritatively, to wider public notice. You may think that allowing the interview, which apparently does not require your specialist knowledge, could take up a lot of time, and that you have better things to do than offer a primer on viruses to someone who would rather be covering a football match. Nevertheless, it is a useful exercise for you to go back to basics once in a while, and since the reporter has to write only five hundred words by tomorrow he won't want to hang around. If you really are too busy, suggest a colleague who you think might be happy to help.

Beware: reporters have a way of coping with uncooperative interviewees, particularly if the story is controversial or is directly concerned with the interviewee or their work. Either the reporters pester them until they give in, or they write the story without the interview, and state plainly that, after repeated attempts, the interviewee declined to respond to questions. If you have recently published in a widely read journal, or if your institution has issued a press release, you have in effect invited journalists to contact you, so co-operate when they do. If you are reluctant to speak because the information is under review, explain the situation to the journalist and offer to get in touch when the work is published.

How about this situation? The telephone rings: 'Hello, Dr Barbara Smith? My name is Felicity Harper; I'm doing an article for the *Florida Gazette* on recent developments in space technology. From your statement in *Science* a couple of months ago I see that you disagree with Professor John Harry and others about the suitability of this new heat-resistant ceramic cladding. I am anxious to get both sides of this story. Can we talk further, at your convenience?'

This is more promising. You have some assurance that the reporter has done some preliminary preparation, and you have some idea of what the *Florida Gazette* is all about. It sounds like Ms Harper has chosen the right person and that you can supply the information she needs, but it's worth checking all the same. Next, ask about the journalist's background (she may be a science graduate, which would make your job easier), about the proposed article, and about the kind of interview that's required. While you have differences with some of your colleagues, you do not want these to be exaggerated at a time when there is a furious public debate about the future of the space programme. However, a few minutes of general conversation will reassure you that you can trust Ms Harper, and that the interview will turn out well.

If you have published something controversial or consequential, or if you have won an award, or if the subject of your work is suddenly newsworthy, be prepared for media attention. Sort out your notes, collect your thoughts, and discuss the material with your colleagues. Keep some time free, and if you have to go away find someone who can deal with the journalists in your absence, or you may find that the interview is given by someone who is not qualified to do so. You could pre-empt a media onslaught by choosing an appropriate journalist whose work you know and respect, and offering him or her your story. You will then have more control over what happens next.

Rat race

Scientists are to journalists what rats are to scientists.

Victor Cohn, medical writer, *Washington Post*

Preparing for an interview

Whenever you are asked for an interview you can check out your interviewer by doing some interviewing yourself. You want satisfactory answers to these questions:

The interviewer

What is his or her job (news reporter, feature writer, researcher . . .), and what is his or her background?

The publication

If you are not familiar with the publication, get hold of a copy or ask the interviewer to bring one. Don't dismiss local papers – they have a large and important readership whose support and goodwill may be vital to your institution. Local editors have a good sense of what makes a national story and may be able to circulate your information more widely than you might expect. Ask about readership and circulation, and whether the newspaper or magazine carries regular features on science.

Why me?

It is not always easy for a non-scientist to distinguish between fields, and you may not be the right person for the job. How did the reporter obtain your name? Has he or she spoken to any of your colleagues yet? If so, whom? They may be able to tell you about the journalist's manner and style.

The scope of the interview

What precisely is the article to be about? How long will it be? The interviewer might be able to let you have written questions in advance, so that you can prepare thoroughly.

The purpose of the interview

What is the interview for? Do they want pictures? Will the information you provide be edited, rewritten or summarized, or will it be printed with only minor changes? Might the interviewer be able to let you see final version of the article before it is published?

Where and how?

There are many types of interview. Will it be business-like, in your office? Informal, over lunch? Hurried, over the phone?

Some people feel that it is better to go unprepared into an interview, and that this results in more spontaneous and natural answers. However, if you are nervous it may help to anticipate some of the questions and to prepare brief answers to them. If you have time, write down what you are going to say. Think about what questions you might be asked (including the difficult ones), and think about how you might answer them. When you write, think about whether the language you are using will be appropriate for both the reporter and the readers of the final article. It is often difficult to think of a clear, accurate and jargon-free explanation on the spur of the moment, and writing will help you to get it clear in your mind. It will also give you an idea of the length of what you are planning to say, and you should compare this with the expected length of the article. If you write more than twice the number of words that the reporter wants, you are including too much detail. Try to work on the same scale as the reporter – the article is then more likely to be an accurate representation of what you said. However, writing the information down is not acceptable substitute for talking about it: the journalist is coming to talk to you, not to sit in your office and read a paper. Put the written version aside during the interview, and only refer to it if you think you've left something out. You could offer the journalist a copy to take away, which might help prevent errors in the published version. If you have a lot of data or some good pictures you could include them in the written version, and point them out to your interviewer.

Interviews conducted over the telephone save a lot of time, but don't let an unexpected call surprise you into hasty answers. There are few occasions when fifteen minutes make any difference, and this will give you a chance to chat to colleagues and collect your thoughts. Always return calls promptly, and if you have asked the journalist to ring you back, keep your line clear and stay by the phone.

Call right back

> Calling the reporter back the next day is often tantamount to not returning the call.
>
> Carol L. Rogers, Office of Communications,
> American Association for the Advancement of Science

The main event

The journalist who comes to interview you may have no science training and few scientific contacts, and will be in a hurry. He or she may be worried about appearing ignorant in front of a specialist. The more at ease you are during the interview, and the more apposite your answers, the more successful the interview will be. Make friends with your reporter – it will make the interview go smoothly, and you will get a much better deal when the article is written. If the reporter likes you, he or she will assume that the readers will as well.

Do you want the reporter to meet a white coat surrounded by books, a suit at a computer or a pair of jeans taking a stroll in the garden? Wherever you hold your interview, do your best to prevent interruptions from the phone or visitors. It is probably better not to have your press officer sitting in on the interview, as this will make it look as though you are expecting trouble or have something to hide. Arrange for someone to be close at hand to help you if you need something photocopied or fetched from the library. Talk with your colleagues before giving the interview. You could suggest to the interviewer that one of your colleagues could do the interview with you – you may give a livelier and more balanced account together, and, if you listen to each other, a more complete one. You will also avoid being depicted as the stereotypical lonely, dedicated, single-minded scientist, and will show instead that science is a co-operative and human enterprise. Don't be surprised, though, if the interviewer wants just you – they may not like being outnumbered. If you do the interview alone but are talking about work you did with six colleagues, don't expect the interviewer to be captivated by your list of their names. One name is 'human interest'; seven are very boring indeed.

The most effective way to respond to an interview question is to get to the point and stick to it. Steer clear of jargon and detail – sort out what's important and limit yourself to a few essential pieces of data. Use ordinary conversational language: words like 'consider', 'because' and 'if' are better than 'give consideration to', 'due to the fact that', and 'in the event that'. The more diversions you make, the more cutting will have to be done, and the more likely it is that some of the important bits will be edited out as well. Be direct. Don't say 'all the evidence would seem to indicate that this claim is not justified' when you could say 'that's not right'. Use examples, and make the relationship between ideas clear. Emphasize your main points: you can even say 'and this is the important point . . . '. Do as much interpretation as you can, so that what appears in print is your version and not the journalist's translation. If you can think of any lively illustrations or anecdotes or analogies be sure to use them. No journalist ever thought badly of a scientist for being articulate and amusing.

Wily politicians occasionally talk 'off the record'; they give background information or gossip that may not be reported directly or make non-attributable statements, which means that the information can be used without naming its source. It is unlikely that you will need to resort to either of these tactics, but if your information is confidential or sensitive, think very carefully before presenting it in any form to a reporter whose job it is, after all, to report news. It's easy, during a relaxed chat with a reporter, to say things you would not want recorded, so take care – if you don't want something to appear in print, don't say it.

On my way to the lab today . . .

Avoid offhand comments. Do not make jokes or tactless remarks; you may well find them quoted. Any remark which begins 'Don't quote me on this but . . . ' will almost certainly find its way into the story.

Frank Albrighton, information officer,
University of Birmingham

If you don't know the answer to a question don't bluff – the journalist will be able to tell. If it is something no one knows ('how long until fusion power will be available commercially?'), explain why you can't give an answer. If it's simply something outside your field, suggest someone who might be able to help. If you don't understand the question, say so. If you have been asked a question which you think is irrelevant, answer it as briefly as you can and try to guide the conversation back to what's important. Be polite, however ridiculous the question might seem, and remember that the reporter is asking questions that may have nothing to do with his personal opinions, on behalf of the general public. If you feel that a vital aspect of the subject has not been covered, say so.

Applications and implications are what interest the general public the most, and if you won't discuss them the reporter may speculate, perhaps wildly. It's important to discuss the limitations, uncertainties and consequences of your work if you get the chance. You will need to qualify your sentences, rather than add qualifiers at the end of your interview which may be chopped. Be moderate: do you mean 'a major leap forward', or is it really just 'a step in the right direction'? It is better to say 'this method is promising' rather than 'we are all very excited about the potential of this method', and to talk of 'treatments' rather than 'cures'. Over-optimistic vocabulary is particularly dangerous in medical news, since patients may then turn up at their doctor's surgery clutching a newspaper article long before the doctor has had a chance to study the research. But make sure you say something – it's easy to qualify a statement to such an extent that it ends up sounding as though nothing has actually happened. Earlier we mentioned that the reporter covering Rutherford's talk about his new atomic structure missed a good story: it could well have been because Rutherford ended his lecture with this masterpiece of moderation: 'we are on the threshold of an enquiry which might lead to more definite knowledge of atomic structure.'

Words and values

> Was Three Mile Island an 'accident' or an 'incident'? Was Chernobyl a 'disaster' or an 'event'? Is dioxin a 'doomsday chemical' or a 'potential risk'? . . . Some words imply disorder or chaos; others certainty and scientific precision. Selective use of adjectives can trivialize an event or render it important; marginalize some groups, empower others; define an issue as a problem or reduce it to routine.
>
> Dorothy Nelkin, media and science policy analyst

When the reporter is not quoting you directly, they will be trying to achieve a compromise between what you said in your own words and what they must write in theirs. They'll have plenty of journalistic tricks for doing this, but even so you will stand more of a chance if you can edit yourself. This means you should be as succinct as possible. Few reporters are satisfied with taking what you've said and transcribing it word for word for their eager readers, however pointed your remarks. Reporters have their own individual style, and will want to give the article an original angle. If you give them a portrait in black and white, they'll add a dash of colour. You'll find that a routine remark was uttered 'with passion' or that your clear exposition was 'enough to give Einstein a migraine'. Then the editor may change the article a little (or a lot), so if you and your comments emerge from the process in recognizable shape, everyone has done his job properly and you can be justly proud that your patience and preparation have been rewarded.

Anyone who has ever given an interview – even those who have given many – will know that the time when the perfect answer springs most readily to mind is when the reporter has just left. If this happens to you, telephone the reporter as soon as you can and ask to have the choice morsel recorded.

The most common complaint scientists make about interviews is that a lot of what they've said is omitted from the published version. Realize that you will never be reported fully, just as only selected parts of a politician's speech are reported – only a very small proportion of what you have said will survive the editing process. Although it isn't always possible, you may be able to get hold of a copy of the article before it is published, and then you can correct any errors. Your input at this stage is seen by some as a form of censorship, as if you are trying to control what other people think and read. If journalists are reluctant to allow this it will be because they feel that an article which goes out under their name should reflect what they feel is important about the interview. You should bear this in mind when you ask to see the article, and if you are allowed to check it, you should remember that you have no right to comment on the style or attitude of the piece, and should confine yourself to judging its factual content. If the schedule is tight you could ask for the article to be read to you over the phone or faxed to you. You can then amend it and return it right away. Remember that no one is under any obligation to take any notice of your comments.

Hide and Seek

> Being interviewed poses a substantial element of hide-and-seek between a reporter looking for juicy quotes and a scientist trying to minimize danger to his reputation from oversimple public pronouncements or outright media distortion.
>
> Stephen H. Schneider, climatologist and broadcaster

Information services

Instead of waiting to be called out of the blue, you can increase your contact with journalists by giving your name to a 'media resource service'. These organizations keep registers of scientists who are willing and able to provide journalists with background information and advice. The service is free to all concerned. There are media resource services in Britain and the USA, and if they haven't got in touch with you, you can contact them and offer to join the register. Their addresses are given on page 174. Once you are on the list you will receive calls from them asking your permission to pass on your name and number to a journalist, so only volunteer if you are accessible and prepared to respond quickly. If when the phone rings you can't help, you may be able to suggest a colleague who can.

Your own institution may have a public relations or information office of some kind. These often function extremely well as a bridge between the research and media communities, and actively solicit media interest in the scientific activities of their institutions by issuing regular news sheets and press releases. Local and national newspapers call them for updates and information. In a well-organized set-up you will be contacted when your work becomes newsworthy. If you aren't, you should approach the appropriate officer, perhaps through your department or laboratory head, to explain that you are interested in having your work communicated to the public. The office should work as an active intermediary for you, inviting media attention and providing background information on your work. For your part, you should give a brief account of your career to the appropriate officer, and keep him or her up to date with your work. The office will cultivate press contacts by post and phone, and maintain a clippings file of published reports about your research. Universities, colleges and institutes are more aware than ever before of the importance of their public profile, and you should find a skilled and committed team in your public relations or information office. Although this team may not include anyone with scientific qualifications, their experience of publicity, journalism and public relations will be a great help to you in your dealings with journalists.

Aftermath

Once your work has been reported, you may get follow-up enquiries from colleagues and the public. Anticipate likely questions and prepare answers. Find some literature that you can recommend as further reading. If you are overwhelmed by telephone calls, put aside part of the day to answer them, and set up a recorded message to let people know when they can call. You could brief a student or colleague to answer the calls for you, or get someone to take down names and addresses and send out a reading list or summary. The fuss will die down sooner than you think.

Trainee reporters receive instruction on 'reportorial ethics' and on how to avoid bias and prejudice. They are taught to be open-minded and objective, and to report the truth so far as they are able. Then they join a newspaper owned by a corporation with many vested interests. On occasion, a newspaper will mount a tenacious campaign to publicize an injustice even when this seems to run against its own corporate interests – the sustained reporting by the London *Sunday Times* of the thalidomide scandal in the early 1970s is an example of committed reporting at its best – but usually the vested interests are a stern guide of editorial policy. It's true that journalistic freedom still exists, but it's a precious bloom fighting for survival in a harsh climate of publishing-house takeovers, competition for readers and a cut-throat struggle for advertising revenue. On top of that, the journalist has to conform to the style of the publication, fit the space available, please the editor and interest the readers. Put a kind-hearted, free-thinking young reporter into a typical newspaper office for a few months and you'll quickly see what the phrase 'conflict of interest' means.

This description may be a caricature, but it will give you some idea of the constraints under which journalists work. You should bear these constraints in mind if you are considering complaining about an article. You should also think about why the errors arose: could it have been your fault? If the reporter did not understand you it may have been because you failed to make yourself understood. Chalk it up to experience and be more careful next time. Some feedback can help both you and your interviewer; if you liked the article, say so.

Save your complaints for those rare errors which really matter. First, talk to colleagues – you may be getting steamed up over something which they think is trivial. Will the public – the readers – notice the error? Does it make any real difference to the tone or message of the article? Will anyone remember the article anyway? As biologist Paul Ehrlich realized, 'there is nothing so dead as yesterday's news; no matter how hideously garbled it is, it's all over'. If you still feel you must complain, speak to your interviewer directly in the first instance. He or she can ignore you, apologize, arrange for a correction to be published (which people may or may not notice), or do a follow-up article to make up for omissions. If you fail to receive redress, your next step should be to contact the editor in charge with a private or 'for publication' letter. If that strategy does not succeed, and you feel that your case is important enough and strong enough to merit a legal action, you should discuss the matter with your colleagues, sponsors and employers, and, if you have their support, contact a lawyer or citizens' advice office for help. You may find that your own fight for justice finds many supporters. On the other hand, you may find in the future that the newspaper you have challenged will have even less time for science and scientists.

Recommended reading

John Brady, 1976, *The Craft of Interviewing* (New York: Random House). A manual for journalists on how to ask the question that elicits the right answer. Useful for interviewees, because forewarned is forearmed.

4.2 Television

Television today occupies a place in our lives which would have astonished the early pioneers who, with their vacuum tubes and valves, made broadcasting possible. In industrialized countries television sets are installed in 95% of homes; many households have two sets or more. Television watching, at twenty-five hours per person per week on average, is the most popular leisure occupation in the western world.

Television has great reach and authority, and its strength is its ability to take information, presented in a direct and accessible way, right into people's homes. A forty-minute documentary will contain more information than a feature in *New Scientist*, more pictures than several editions of *Na-*

ture, and will reach many times more people than either of these. Your boss may not watch your programme, but your neighbour might. Your colleagues may turn up their noses at it, but the people who live near your laboratory may feel happier about what goes on there, and the others will realize that science is done by people, and that it is part of their lives.

The right stuff

> Television is an admirable medium for expression in several ways: powerful and immediate to the eye, able to take the spectator bodily into the places and processes that are described, and conversational enough to make him conscious that what he witnesses are not events but actual people.
>
> Jacob Bronowski, biologist and broadcaster

We know that people are influenced by what they see on television, and that they can be induced to buy new brands of cornflakes, to vote for parties and presidents, to part with their money and to acquire new opinions about a whole range of issues. Science on television has certainly been a major factor in the development of the public's perception of scientists and science. One important consequence of television science is that it captures the interest of children – even those too young to read are captivated by the astounding images of the natural world which television brings into our homes. Ask around at work and you will find that a whole new generation of scientists has emerged from among those children whose first sweet taste of nature was had not in the library or the schoolroom, but from television, which took them to the plains of Africa and under great oceans whose quiet corners proved more startling than the Moon.

The power of television is one which scientists should be happy to use, and yet there is some snobbery towards it. Some people believe that it is impossible to be 'serious' on television, and that the medium inevitably trivializes everything it transmits. The novelist Graham Greene is no fan of television, and is extremely reluctant to appear on it. 'If one is bad', he said, 'why do it? And if one is good, one ceases to be a writer and becomes a comedian.' Others, while believing in the benefits of television, worry that they may be compromised simply because this attitude persists among their peers. However, while it may have begun its life as a simple purveyor of light entertainment, television has attained a sophistication few would have predicted. It has matured quickly, and reflects contemporary life and ideas with a swiftness and clarity that the other arts rarely achieve.

Audiences too have matured. People can differentiate between information and entertainment, even if both come in the same breath. Because they can differentiate they can also chose, and some may opt to miss science feature programmes in favour of 'light entertainment', a term which covers everything from chat shows to circuses. This means that if you turn down an invitation to be interviewed by the personality of the time, you will miss a chance to address an audience that wouldn't usually pay much attention to science. The fact that you have been invited to appear on the programme means that the producers, who know their audiences well, believe that your story and your manner are interesting enough to hold the attention of an audience that expects entertainment. If you come across as a comedian, the responsibility will rest entirely with you. Television and its audiences can and should be taken seriously, and in return will offer you the chance to bring the sights and sounds of science into the homes of millions.

Taking to the air

Most science on television is chosen by the production staff, who find their stories in science magazines and academic journals, in the news, by following up their own ideas and interests, and by talking with scientists. If a producer or researcher is interested in your story and wants to know more, he or she will contact you. Deal with this as you would with a newspaper interview. However, when a television company calls you, it may not simply be for information. It is expensive to send a film crew to your office, so a phone-call may also be your audition. You may be a useful source of information, but you may also be a useful source of pictures and a useful voice. If you sound enthusiastic and can provide information clearly and in a style appropriate to the programme, you may well find yourself on television.

If you think your work would be of interest to the viewing public, chose a suitable programme and contact the producer. If your idea is not one that needs to be told immediately, write a proposal along the lines of a book proposal. If your story needs to be told quickly, telephone the production office and be ready to discuss your idea at length. You will need to have thought the idea through thoroughly beforehand. Ask yourself the four questions on page 33, and then prepare as you would for an interview. The extra vital ingredient that you will have to supply along with your story is pictures. If your story will take up three minutes in a magazine programme, you will need to describe three minutes of pictures. For a forty-minute documentary you will have to supply forty minutes of pictures. This is often the most difficult aspect of putting science on television, and many a good science story doesn't make it to the screen simply because it has no pictures.

A television programme is not always the ideal form for presenting scientific ideas. If your subject won't make sense unless you pile in lots of details and data, it won't make good television – the more accurate you try to be, the more overwhelmed your audience will feel. Most people can only take in one or two new ideas in any one programme, so your proposal should consist of a very small number of ideas, perhaps presented in different ways and with a lot of examples.

TV imperatives

> There's a 'complementary pair' in television called 'Accuracy and Clarity'. We can make something very clear as long as we don't worry too much about accuracy. On the other hand, if we go for total accuracy, we make it unclear in this medium. It's unclear because television demands minimal explanations. As soon as one tries to get very accurate, clarity slips beyond the viewer's grasp. Television demands fast explanations.
>
> Gerald F. Wheeler, physicist and TV producer

If you have no luck with the major television companies, try the independents. There are now over six hundred independent production companies in the UK alone, and some of these specialize in science documentaries. Independent companies are happier to work with independent scientists than they are to work for companies, so if you have an idea you feel would make a good programme you should not be dissuaded by the idea that only big organizations have any power in the media. A big organization may sponsor your work on a script or proposal, but if your idea is accepted the production or broadcasting company, for reasons of copyright and integrity, will fund the programme themselves. An independent company will not produce your film unless they have found a broadcaster for it, as documentaries are expensive to make – you won't get 40 minutes for much less than £70 000.

Accepting invitations

When you write an article for a newspaper your work will be judged by the standards which apply to professional journalists. The broadcast media are easier on the amateur participant – one of their strengths is that they provide a platform for people from every walk of life. A sizeable fraction of broadcast material relies on the contribution of amateurs: news reports, documentary interviews, access programmes, studio debates, chat shows, game shows, sports coverage and outside broadcasts all rely on contributions from people who are not professional broadcasters. This means that the programme makers welcome – and need – amateurs, and will make your participation as easy for you as they can. Provided you acknowledge the constraints of the medium and recognize the producer or editor's responsibility for the final product, you will be treated as a valuable and very welcome guest.

When you are invited to appear on television you should ask yourself, and indeed the person who invited you, why you have been chosen. Are you the best person for the job? If you can think of someone better qualified, tell the producer in good time, although you may find that you have been asked because they have already interviewed Dr X and found him monotonous and twitchy. If you are not interested in the subject don't go – they need an enthusiast. If you are representing a pressure group or institution, or if the producer thinks you are, be sure you know exactly what line you'll be expected to present. If this doesn't agree with your personal opinion, then don't go – you will find yourself spending the whole interview denying and qualifying and nothing will get said.

Before you start to prepare for the programme, watch it and talk to people who've seen it. You need to know about the set, how long you are likely to appear for, and what the general atmosphere of that programme is likely to be. Find out who watches the programme, and what they might be expected to know about your subject. The production team may have prepared an introductory film or preamble. Find out what's in this, so that you can follow on rather than repeat. On the other hand, they may expect you to provide the material for your own introduction, so gather up some historical background or a case history. If you are contributing to a programme where several people will appear, make sure that you are clear about the subject being discussed – it may overlap with only a small area of your work. Find out what's going to be in the rest of the programme. You may be able to link your item to one of the others. When you are invited to broadcast it will be to talk about a specific subject, and you should not take advantage of your time on the air to talk about any subject other than the matter in hand.

Don't worry too much about your appearance, but some attention to dress can be worthwhile. Try to avoid unconsciously reinforcing stereotypes: dress as yourself, not as 'the scientist'. Studios can be hot, so don't wear too much. Black and white don't look good on television, and you should avoid big patterns, checks and stripes. Avoid frills and asymmetrical details, and make sure that your clothes look good when you sit down as well as when you are standing up. Rely on the studio for make-up, and don't change it afterwards. It may look peculiar to you but it will look just right on television.

In the studio

Television has long since stopped trying to conceal its technology, and every now and then the viewers are treated to a panoramic view of the cameras, technicians and lights. Even though most people now know what a television studio looks like, it's still easy to be overawed by it – just as media people often are if they come to see you in your laboratory. At first, you'll think it incredible that people work willingly in such surroundings, but by the time you've finished you'll probably be ready for another invitation. The television studio, with all its paraphernalia, has an allure and a magic of its own.

Apart from your fellow guests and presenter, there are four key people involved in making a television programme. The producer is responsible for the practical process of making the programme, and was probably responsible for inviting you to take part. The editor, who may have helped choose you, will have decided on the format and content of the programme. The director sits in the gallery and controls the sound and vision, and the floor manager, with whom you are likely to have some contact, makes sure that the people and equipment are in the right places. The programme is in their hands – not yours – so don't worry about the technical side of the process: concentrate on answering the questions.

When you get to the studio, ask if you can see the set. Practice any steps, and test your chair for comfort and squeaks. Find out where the microphone is and what its range is. The technicians will have put it in the right place, so don't lean towards it, and if it's attached to you don't fiddle with it. If the microphone is uncomfortable, the light is in your eyes or you are desperate for a glass of water, say so.

The interview

Most of the advice about press interviews applies to broadcast interviews, but there are some differences. One is that broadcasting works to even tighter schedules than newspapers. Your invitation to the studio will in all likelihood be to discuss a piece of news, and if you get twenty-four hours' notice you can count yourself lucky: many get only a few minutes. You may get a call at 11.30, be recorded at 12.15 and be broadcast on the lunchtime bulletin.

If you are asked to contribute to a programme, ask the same questions as you would ask before giving a press interview. Think about the programme and its reputation. Will the interview be recorded or broadcast live? What is the programme about? What questions should you expect?

Most interviews are recorded, but even then they can be intimidating and call for as much confidence on your part as you can muster. Set up a practice interview beforehand if you have time, or pretend to be the interviewer and ask yourself questions – those you expect, and those you hope you won't be asked. Do not presume that your interviewer will agree with you; expect to have to defend your position.

If the film crew and reporter come to visit you, you can get the day off to a good start by making their life a little easier. Make sure that there are enough spaces reserved for them in the car-park, and that someone is on hand to meet them. See that your visitors are comfortable, and tell your colleagues who they are and what they are doing, so that you are not interrupted too often. Get your phone-calls intercepted. Television crews seem to prefer their subject to be sitting in front of shelves of books and wearing a white coat, but neither of these features is essential. A white coat can intimidate or mislead – the public will think that you are either a doctor or do questionable experiments on animals. You could suggest that the interview is held in your lab or common room, or in the grounds of your institution.

If you are travelling to the interview, allow yourself plenty of time to miss trains, change a tyre, get lost, and be held up by the studio security. If you are nervous, take along a friend who can drive you, park the car, check your clothes and find coffee. Ask the producer all the questions you want to ask, even if they are trivial. You could ask about any technical words you intend to use – producers have a good idea of which words need defining and which will be understood straight away.

Television is an intimate medium – your face is in people's homes, and you should behave as though you are. There is no need to give a performance, and ordinary spoken English is the most appropriate. If you use jargon the interviewer will ask you to explain yourself, which will waste time, and you will never be invited back. Remember that your audience is a group of people, not a cross section or rating. Keep those people in mind throughout – the interviewer is asking questions on their behalf, not his own.

You and the viewer

> Audiences may resent being treated like school-children, although they enjoy having their curiosity aroused and satisfied, as a good teacher would do for them. They like speakers who wear their authority lightly. They also prefer to be regarded as intelligent people who can make up their own minds about clearly presented evidence.
>
> Alec Nisbett, science producer, BBC TV

You will probably meet the interviewer before the recording, but don't expect to talk about the subject of the interview. The interview should sound spontaneous, and it will help if you are surprised by some of the questions and the interviewer is surprised by the answers. If you do discuss the questions beforehand, remember when you get to the real thing that the audience won't know what you've said off the air, and so you should answer as though you are being asked the questions for the first time.

Although documentaries are the most obvious manifestation of science broadcasting, the vast majority of scientists on television and radio are on the news. If you are being interviewed for a news programme, be prepared to make your point in one or two concise sentences. Get to the important information as early as possible. The conclusion should come at the start, like in a news story; that way, if the interview is cut short you will have made your point. Do as much editing for yourself as you can – avoid digressing and being irrelevant. Don't worry about creating a background or context: this should have been done for you. If you embark on a flowery preamble you will be cut off before you get to the point. You may find that there are fifteen seconds to go and the interviewer turns to you and asks: 'Dr Jones, could you tell me in a few words what this fuss is really all about?' Instead of mumbling a few words, have ready a clear and positive statement with which to finish.

The philosopher and mathematician Bertrand Russell was not a man often left speechless, but he was caught out on at least one occasion. One day in a taxi the driver turned to him and said, 'Aren't you the philosopher Bertrand Russell?' When Russell replied that he was, the driver said, 'Tell me then, what's it all about?' This simple question floored Russell and left him puzzling about the implications of a lifetime's work. However, Russell's interviewer was a cab-driver, and would not have made much of a living out of asking questions. The interviewer's art is to ask questions which elicit the required response. This is why there is no need to be overawed by famous interviewers: they get to be famous by being good at their job, and so will make things easy for you. You will not be asked trick questions. If you are not sure about the question, answer what you think it means. If you do get the wrong end of the stick, the interviewer will realize his or her error and will put you right by asking a more specific question. If you want to come into a group discussion, indicate to the interviewer, and don't be afraid to argue.

If you need time to think, take it. A couple of seconds can make all the difference to your answer and the audience won't notice the pause. In a recording the silence may even be edited out. If you stop for too long the interviewer will help you by asking a supplementary question. Answer the question firmly, but don't try to keep talking if the interviewer is trying to shut you up. On the other hand, a simple 'yes' or 'no' rarely constitutes an adequate answer.

Avoid prefacing every statement with 'I think' unless you are going to discuss other people's opinions as well as your own. The audience will be aware that what they are hearing is your opinion. Don't say 'I'm glad you asked me that question' or 'now this is very interesting' – statements like these are not interesting and waste time. There is no need to take sheets of data with you – you will not be asked for specific information. Words like 'increase', 'level out', 'significant improvement', 'serious problem' and 'good news' are all better than '32.57 plus or minus 4.75', and so is 'inconclusive'.

Try to sound enthusiastic and interested in your subject. If you don't, your interview will count for nothing. Even if people listen to you, they won't remember a word you've said.

Dulled to death

> If you are dull, no matter how responsible, or how accurate, or how unbiased, or how complete the rest of what you have to say is – if no one is listening, it doesn't matter. It simply doesn't matter.
>
> Sam Donaldson, ABC News

Try not to give the impression that it's impossible to give an adequate answer in the time available, or that the subject is 'too complicated'. Prepare an outline, or the results without the theory. Be careful not to highlight the ignorance of other panelists or of the interviewer. If you are in dispute with the other scientists, a television studio is certainly not your field of battle. It is better to say 'but the way I see it . . .' than 'Mr X is talking nonsense'. Don't let group discussions become too personal – until such time as you become a 'personality', the audience is only interested in your knowledge and opinions. Even if you are old friends, don't refer to the other panelists by their first names if the audience has only been given surnames.

Do not produce surprise objects in mid-interview – check with the producer and interviewer first. The technicians will need to be told if a close-up is required, and your object may be too shiny or too small. If you do take objects along, make sure that they are relevant and useful. A bottle of colourless liquid will cause more consternation than enlightenment if you've taken it along to show the audience that this lethal chemical looks just like water.

Talk to the interviewer rather than to the camera, and try to sit still. You will often be in close-up, and fidgeting, hand waving and finger waggling which might be tolerable in a lecture theatre will be horribly magnified by the television camera. Try not to nod or grunt your approval or otherwise while the interviewer is speaking, and don't interrupt. Listen to the question, and answer the question that is actually asked, even if it's not the question you were expecting. If you say 'what you really mean to ask me is . . .' you will look as though you are trying to avoid the question.

Interviewers find that the most difficult thing to do is what they actually appear to be doing – listening to what you have to say. Occasionally there may be a pause between your answer and their next question, or a quizzical look on the interviewer's face. It may seem as if your interviewer has totally lost track, but don't panic: let them retrieve the situation. Interviewers have 101 things to think about apart from their conversation with you – the floor manager's signals, the studio clock, their own commentary, and so on – and sometimes they do lose track. That, however, is their problem. You just have to answer the questions.

If studio discussions seem to be getting bogged down or, worse still, if everyone seems to be reaching agreement, interviewers may try to heat things up a little. They may contradict you or exaggerate your viewpoint in order to provoke a spirited response. 'Surely, Dr Williams, some might accuse you of . . .'; 'what is your answer to the charge that . . .'; 'is it really true that . . .?' Words like 'accuse', 'charge', and 'really' are intentionally incendiary. You can respond emphatically, but don't get angry or aggressive. Don't be defensive either, or the interviewer will assume that you've got something to hide and will try to find out what it is.

Television news is broadcast in simple and dramatic terms and with the minimum of equivocation. If you are quoted in a broadcast you should expect your sentences and paragraphs to be chopped up and rearranged, so make sure that any qualifications you make are integral to your statements, rather than added on at the end – additions can easily be subtracted.

Very few programmes are broadcast live. If yours is, think very hard before accepting the invitation – it is particularly important that you are completely sure of your subject and are confident enough to talk about any aspect of it. What worries people who do live broadcasts is the idea that millions of people are watching them. The best way to deal with this is to think of the viewers as eavesdroppers, and to concentrate on the private conversation you are having in a studio with a few people who are representatives of the public.

Tell your colleagues that you are going to be on television and ask them for comments afterwards, and video your programme and watch it yourself. This will help your performance next time. As a professional scientist you are an amateur in the media, so be prepared to learn from the professionals. Remember that all broadcasters' primary responsibility is to their audience. Whenever you broadcast, the public will think of you as 'the scientist'. You are representing your institution and your profession, and by taking on this responsibility seriously you can contribute to improving the relationship between science and the public.

Recommended reading

Evan Blythin and Larry A. Samovar, 1985, *Communicating Effectively on TV* (Belmont, CA: Wadsworth). Helpful advice on preparation and performance.

Eric Paice, 1984, *The Way to Write for Television* (North Pomfret, VT: David & Charles). This covers fiction as well as non-fiction, and offers advice on the marketing and presentation of scripts.

4.3 Radio

The power of radio was demonstrated beyond question by Orson Welles's 1938 broadcast of *War of the Worlds*. A million Americans thought that they were listening to a live account of a real Martian invasion, and people panicked, running screaming through the streets, sealing their homes against gas attacks, or fleeing to underground shelters. They believed that our planet was being invaded because *War of the Worlds* was broadcast in the style of a news report, and people trusted the great wooden wireless sets that filled their evenings with news and music.

Radio delivers fact as forcefully as it does fiction. Science doesn't constitute a large fraction of radio broadcasting but, as with television, science is covered in news bulletins and documentaries. American radio only occasionally features science, but in Britain BBC Radio 4 devotes a valuable thirty minutes to science every week. The BBC World Service broadcasts about two hours of science every week, to a world-wide audience of tens of millions. The BBC's radio science unit maintains a steady, high quality output which is widely respected, particularly by people in authority who have little time for magazines or television, and who rely on the crisp erudition of Radio 4 to keep them informed. The influence of radio science is therefore far greater than might be expected from its audience figures, and this, together with the unfailing competence of its practitioners and the opportunities it offers the amateur broadcaster, means that radio is a medium worth taking very seriously indeed.

Radio is different from television: it has no pictures. This apparent disadvantage of radio has, in a curious way, become radio's strength. Because producers and presenters have had access to few of the resources of their colleagues in television, they have been thrown back on their own ingenuity, creating pictures in the minds of their listeners. Compared to television, radio is a frail and non-compulsive medium: a voice which emerges from an invisible speaker. Yet while television has become technically sophisticated and complex, and its programmes the work of large teams, radio has remained a personal medium. The speaker addresses the listener directly, without ceremony and without fuss.

Radio producers find their stories mostly by keeping an eye on the specialist press. Sometimes people contact them and tell them about their work, or maybe a scientist sees something happening in his or her field and reads up on it and offers to talk about it. Some scientists are so good at finding stories that they have become regulars on British radio. This doesn't mean that producers are lazy and rely on the same old voices – their problem is that new voices don't often get in touch. Some radio stations, especially local ones, don't feature science simply because they have no scientific contacts, so may be very pleased to hear from you. If you want to be on the radio and you have a story which you think would interest the public, telephone the news or features production office at your local station. If you have news, be prepared to give the story over the phone in two minutes. For a feature you could have five minutes. If the producer or reporter is interested in the story they will telephone you. This phone-call is effectively an audition: they will want to hear how you talk about your work. If they like the sound of you they will set up a recording session, which will either be an interview at your place of work, or an interview in a recording studio near you.

If the interviewer comes to you, your interview will be a friendly chat in familiar surroundings. You will feel comfortable – in theory at least – and the reporter will get an idea of your working environment. He or she might record some background noise to provide the listeners with some audible scenery. If you go to a radio studio you may meet a reporter and a producer and find technicians fiddling with equipment, but you may also find yourself all alone in a tiny, over-upholstered but barely furnished room. Since most people don't live near a major radio station you may be invited to a local studio to chat with an interviewer who is hundreds of miles away. You will be on your own: your interviewer's voice will resound from the void and you will reply into the abyss. This is difficult – you are alone in a strange place responding to a voice whose owner you have never met. It's easy to forget that when your interview is broadcast, you will be talking to real people, thousands of them, each in their living room or car. You must imagine that your interview is a conversation not with a disembodied voice, but with a real person who is sitting across the table from you. People find different ways of achieving this – some shut their eyes and imagine their interviewer, or have in mind a friend they can talk to. One scientist copes with the absent interviewer by drawing a face on a piece of paper and talking to that. The experience is rather like having a telephone conversation, so if you are usually coherent and animated on the phone, you'll probably cope well with absentee interviewers. If you listen to the programme before your interview, you'll know how the time is divided between presenters and guests, and how long you'll have to make your point.

Radio people work on a different time scale from everybody else. Programmes are timed down to the last second, and to be 'on time' in radioland means to arrive about fifteen seconds before you are due to go on the air. However, it is vital to be on time for an appointment at a radio studio, and it is far better to be nine hundred seconds early than fifteen minutes late. Even if you aren't broadcasting live, studios are usually very busy and your recording may be booked into a ten-minute slot between Hymn of the Week and Gardening Time. Arrive late, and you may find that an overworked producer expects you to talk about compost.

Speaking for listeners

Unfortunately, a radio audience is not a captive one: your listeners are not sitting in front of their radios with an ear glued to the speaker. They are driving, cooking, or rushing through breakfast. You are not addressing half-wits, but people who may only be giving you half their attention, which amounts to about the same thing. This means that the language you use must be simple and direct. Jargon is out, and convoluted academic phrasing is definitely out. Producers sometimes find that when they chat beforehand their interviewees are clear and interesting to listen to, but once they get into the interview proper they start talking as though they were dictating an academic paper. Talking to a reporter in a studio is exactly the same as talking, via the radio, to ordinary people in everyday situations.

Brief encounter

If you are having a conversation with a stranger on a train and they ask you about your work, you tell them, explaining as carefully as you can. You don't know anything about the stranger – you've no idea whether they know anything at all about your subject – but you still make yourself understood. If scientists could talk on the radio like they would talk to a stranger on a train, they'd be fine.

Peter Evans, BBC Radio

When the BBC radio science team need an example of a scientist who can speak effectively on the air, they cite Harry Rosenberg, from the Clarendon Laboratory in Oxford. In 1988 Dr Rosenberg was asked to talk about the deposition of diamond films. He began:

> Now there was this discovery that you could actually make thin films of diamond – you can actually lay down thin films of diamond by burning gases that contain diamond. And this really is rather surprising, because we are quite used to burning gases and getting carbon down in the form of soot – if you have a smoky yellow flame and you put a piece of glass in it you get smoked glass and the black on the smoked glass is carbon – graphite. And the secret of making this film, instead of being soot or graphite, diamond does now seem to be fairly well established. And what you do is burn a gas containing carbon – any old, almost any old gas – methane is the one that's usually used, but any gas containing carbon – you burn it in the presence of an extra bit of hydrogen at rather a high temperature, and lo and behold: instead of getting a sooty deposit you get down a very fine thin layer of little crystals of diamond.

Written down this doesn't look too good, but if you read it aloud you will see how its colloquial style and spontaneity make it very easy to follow. Harry Rosenberg is thinking as he speaks, and explaining at thinking pace. He starts by telling us what the discovery is, and then tells us why this is surprising. He talks about something we know – getting soot from a flame – and then uses that to describe the new process. There are pictures in his words: the yellow flame and the blackened glass, and the 'very fine thin layer of tiny crystals of diamond'. He doesn't waste words giving us unnecessary numbers: you won't find phrases like 'an extra bit of hydrogen' or 'rather a high temperature' in a scientific journal, but they are fine for radio. Radio broadcasts are for people interested in the processes and applications of science, not for people who want to reproduce results. Anyone who wants to build a diamond machine will have to look elsewhere for the specifications.

Finding the right level of complexity and the appropriate words for a non-scientific audience is especially difficult when you are answering unexpected questions off the top of your head. It is easy, especially if you are nervous, to retreat into the scientist's emergency supply of stock scientific phrases. So prepare for your interview as you would for a press interview – write about your subject in a language you think the listeners would want to hear, and then read your work out loud. Is your explanation in order? Does one point lead logically into the next? Are your words easy to listen to? Now that you have an idea of the sort of language you should use you can throw away your written work. It won't be wasted, as when you arrive at the studio you will be much better equipped to answer the questions. You could get yourself into the right frame of mind before your interview begins by thinking specifically about a non-scientist friend, and then talk during the interview as though you were talking to them. Aristotle didn't do much broadcasting, but with enviable prescience he left this advice for radio speakers: 'Think like a wise person, but talk in the language of the people.'

If you are writing your own script, remember the difference between written and spoken language. Try going through the process for preparing a talk, giving your talk into a tape-recorder and then writing down what you said. It may require a little polishing but it will sound much better on the air than a written text read aloud. Numbers are a problem: '$236 789 150' takes about five seconds to say, and by the time you reach 'dollars' your listener will have forgotten some of the numbers. Round off your figures, and if you refer to a very large or very small number, use a comparison to illustrate the size. Keep your sentences short and straightforward, and remember that you can't rely on punctuation to help the audience through convoluted phrasing – they won't be able to see punctuation, and will be relying entirely on the expression in your voice. Make only a very small number of points, and be sure that your point of view is expressed clearly. Radio, a broadcaster once remarked, is the art of communicating meaning at first hearing.

You and your audience

The personal speaker – you – is the linchpin of spoken broadcasting.

Whether you are reading a script, reviewing a book, or discussing a breakthrough in scientific research, you have to sound as though you are talking to each listener personally, directly and alone.

One to one

> In radio the audience to be aimed at is *an audience of one* (infinitely repeated). And as useful a way as any of drafting a script is to *imagine* that audience of one, pretend you're explaining a point or telling a story to him or her – not immediately, but after you've tried to phrase your message in different ways and settled on the best.
>
> Elwyn Evans, BBC Radio

'Being yourself' on radio isn't easy. After all, you don't normally sit cooped up in a padded cell with only a microphone for company. Just remember that they don't want a character, they don't want a stereotype and they don't want a talking encyclopaedia. They want you. Radio can exert awful punishment on anyone who pretends to be something he is not. If they wanted a performance, they would have hired an actor.

Short-sighted people know that they can't hear too well without their glasses. This is because when we listen to people talking, part of the information we receive is the movement of the speaker's lips and the expression on their face. On radio no one can see your face, so you will sound much less expressive than you usually do. This means that if you are to communicate your enthusiasm you must be more expressive than usual, but make sure that your expression matches the sense and tone of what you are saying. You should be careful not to rush, but there is no need to pronounce words more crisply than usual. The technology is good enough for you to speak in your normal voice and still sound clear. Slips of the tongue are a source of unnecessary worry. If you do make a mistake, do as you would in normal conversation – correct yourself and then carry on. The slip can be edited out of a recording, and if you are live you can console yourself with the fact that interested listeners hardly ever notice mistakes – they have better things to attend to. But it really isn't worth worrying about your speaking voice: if the producer didn't think you could broadcast, he wouldn't have invited you in the first place. Everyone is nervous when they go on the radio, and they worry about small things which no one else notices. Most scientists cope very well with their first broadcast, but all improve with experience. There is no substitute for practice.

Recording your interview will probably take longer than you expect, as only a small proportion of what you say will make it on to the air. The rest of the recording will be discarded, if it's irrelevant, or the information may be used by the presenter to provide the introduction for your interview. Your words may be cut out by the paragraph, by the sentence, or even singly. As with the editing of written words, what you say may have to be cut to make it more straightforward, or to make it fit into the time available. If the programme of which you are a part is to last for 30 minutes it must last for 30 minutes – and that means 1800 seconds, not 1796 or 1811. Pauses are easy to cut, so don't worry about recording a few if you need time to think. BBC Radio's science news programme has been on the air for many years, and the production team have never received a single complaint about their editing. And no one has ever turned down an invitation to broadcast a second time.

Recommended reading

Andrew Boyd, 1988, *Broadcast Journalism: Techniques of Radio and TV News* (Oxford: Heinemann Professional Publishing).

Terry Prone, 1984, *Just a Few Words* (London, New York: Marion Boyars). Terry Prone shares her broadcasting experience in this very readable book, which is subtitled 'everything you need to know if you want to give a lecture, make a speech, appear on television, be interviewed on radio or comment at a meeting'.

Addresses

Organizations promoting science communication

The Royal Society is host to COPUS – the Committee on the Public Understanding of Science. COPUS is a joint venture of the Royal Society, British Association for the Advancement of Science and Royal Institution. It gives the Michael Faraday Award and science book and film prizes, promotes research on science communication and the public understanding of science, gives small grants for work in these fields, and organizes discussion meetings and talks. COPUS welcomes suggestions and is interested to hear of new projects.

COPUS
The Royal Society
6 Carlton House Terrace
London SW1Y 5AG
UK

The British Association for the Advancement for Science organizes training sessions for scientists which are run by media professionals. These 'media workshops' take place across Britain. They give awards for television and film about science, and publish a bi-monthly newsletter 'Science and the Public'. The British Association also organizes Media Fellowships on behalf of COPUS. The fellowships enable scientists to spend 4-8 weeks working alongside professional journalists in different media, and the aim is to create a greater understanding of the media within the scientific community. There are also a number of opportunities for children to become involved in BA projects such as the Creativity in Science and Technology Awards and the Young Investigators' Awards.

Public Affairs Office
British Association for the Advancement of Science
Fortress House
23 Savile Row
London W1X 1AB
UK

American Association for the Advancement of Science includes among its objectives a commitment to 'increase public understanding and appreciation of the importance and promise of the methods of science in human progress, and organizes a number of activities to this end. Its annual meeting is home to gatherings of the National Association of Science Writers; it organizes media fellowships for graduates students; it gives awards to scientists and journalists who have made outstanding contributions to the public understanding of science; it establishes contacts between scientists and their local science museum or science centre; and it sponsors lectures, workshops and symposia designed to foster good relationships between science and the public.

Committee on the Public Understanding of Science and Technology
American Association for the Advancement of Science
1333 H Street, NW
Washington DC 20005
USA

The Association of British Science Writers has over 250 members representing professional science writers in the UK; it holds regular meetings and organizes meetings with policy makers. The Association has produced a guide to careers in science writing, and offers advice and help to writers.

Association of British Science Writers
c/o British Association for the Advancement of Science
Fortress House
23 Savile Row
London W1X 1AB
UK

The National Association of Science Writers has over 1200 members, drawn not only from the media, but also from business and scientific research. It liaises with policy-making bodies and organizes meetings to coincide with the winter meeting of the American Association for the Advancement of Science and with the summer meeting of the American Medical Association. It issues the quarterly *Newsletter* and *Clipsheet*, and a variety of guidebooks; and in 1959 established the Council for the Advancement of Science Writing (see below).

National Association of Science Writers
PO Box 294
Greenlawn
New York 11740
USA

The International Science Writers' Association was formed in 1967 in response to the increasingly international scope of science popularization and technical communications; it serves to provide 'an ever wider circle of contacts.' It issues an occasional *Newsletter*, and has individual membership of over 100 across 21 countries.

International Science Writers' Association
c/o Secretary-Treasurer
7310 Broxbourn Court
Bethesda
Maryland 20817
USA

The Council for the Advancement of Science Writing was established in 1960 by the National Association of Science Writers with a grant from the Alfred P. Sloan Foundation. Its aim is to improve public understanding through more effective use of the media. Activities include on-the-job training schemes for reporters and opportunities for science writers to engage in scientific research in laboratories throughout the USA, reporting workshops, and a fellowship and internship programme. Membership includes writers, editors, scientists and educators.

Council for the Advancement of Science Writing
618 North Elmwood
Oak Park
Illinois 60302
USA

The European Union of Scientific Journalists' Associations was founded in 1971; it produces an occasional newsletter in English which is distributed to European science, technology and medical reporters.

European Union of Scientific Journalists Associations
c/o Dr Ernest Bock
EEC-DG XII
rue de la Loi 200
1049 Brussels
Belgium

Other writers' associations include:

American Medical Writers' Association
9650 Rockville Pike
Bethesda
Maryland 20814
USA

International Society of Medical Writers
c/o Dr Alfred Rottler
Aussere Bayreutherstrasse 72
8500 Nuremberg
Germany

Irish Association of Science & Technology Journalists
303 Martello Estate
Portmarnock
County Dublin
Republic of Ireland

Aviation/Space Writers Association
17 South High Street
Suite 1200
Columbus
Ohio 43215
USA

Illustrators' associations

American Institute of Technical Illustrators Association
2513 Forest Leaf Parkway
Suite 906
Ballwin
Missouri 63011
USA

Association of Medical Illustrators
2692 Hugenot Springs Road
Midlothian
Virginia 23113
USA

Guild of Natural Science Illustrators
PO Box 652
Ben Franklin Station
Washington DC 20044
USA

Society of Engineering Illustrators
c/o Robert A. Clarke
Autodynamics Corporation
30900 Stephenson Highway
Madison Heights
Missouri 48071
USA

Institute of Medical and Biological Illustration
27 Craven St
London WC2 5NX
UK

Institute of Scientific and Technical Communicators
52 Odencroft Road
Britwell
Slough SL2 2BP
UK

Indexers

Society of Indexers
16 Green Road
Birchington,
Kent CT7 9JZ
UK

American Society of Indexers
235 Park Avenue South
New York
New York 10008
USA

Basil Blackwell's *Guide for Authors* is available from bookshops and from:

Basil Blackwell Ltd
108 Cowley Rd
Oxford OX4 1JF
UK

Basil Blackwell Inc.
432 Park Avenue South
Suite 1505
New York
New York 10016
USA

Guides for avoiding gender bias in language

Change Sheet 2
Publication Manual
American Psychological Association
1200 Seventeenth St NW
Washington DC 20036
USA

Prentice-Hall Author's Guide
Prentice-Hall, Inc.
Englewood Cliffs
New Jersey
USA

Wiley Guidelines on Sexism in Language
John Wiley & Sons Ltd
Baffins Lane
Chichester
W Sussex PO19 1UD
UK

John Wiley & Sons, Inc.
605 Third Avenue
New York
New York 10158-0012
USA

Media resource services

The Ciba Foundation
41 Portland Place
London W1N 4BN
UK
Telephone 071 631 1634 or
071 580 0100

Scientists' Institute for Public Information
355 Lexington Avenue
New York
New York 10077
USA
Telephone 800 223 1730 (toll free), or
in New York State 212 661 9110

Sources for the quotations

Page no.	
7	Quoted in Rae Goodell, 1975, *The Visible Scientists* (Boston, Toronto: Little Brown), p. 98.
8	Jean Rostand, 1960, *Science*, **131**, May, 1491.
16	Quoted in Dorothy Nelkin, 1987, *Selling Science* (New York: Freeman), p. 88.
20	Extract taken from *Memories* by Julian Huxley, p. 165, reproduced by kind permission of Unwin Hyman Ltd.
22	Quoted in Barbara Gastel, 1983, *Presenting Science to the Public* (Philadelphia: ISI), p. 24.
39	From correspondence.
46	From correspondence.
48	Quoted in Bill Henderson, 1986, *Rotten Reviews: A Literary Companion* (London: Pushcart Press), p. 91.
52	Bryan Silcock, 1984, *Nature*, **308**, 15 March, 297.
54	J.B.S. Haldane, 1985, How to write a popular scientific article. *On Being the Right Size and Other Essays*, edited by J. Maynard Smith (Oxford: Oxford University Press), p. 161.
59	Ford Madox Ford, 1979, *Memories and Impressions* (London: Penguin), p. 374.
63	William Zinsser, 1988, *On Writing Well* (New York: Harper & Row), p. 25.
68	Robert Graves and Alan Hodge, 1943, *The Reader over your Shoulder* (London: Cape), p. 22.
71	From correspondence.
81	Quoted in Casey Miller and Kate Swift, 1988, *The Handbook of Non-Sexist Writing* (New York: Harper & Row), p. 1.
82	Quoted in Casey Miller and Kate Swift, 1988, *The Handbook of Non-Sexist Writing* (New York: Harper & Row), p. 43.
87	Quoted in Robert A. Day, 1975, How to write a scientific paper. *ASM News*, **41**(7), 486-94, 489.
92	Quoted in Robert Barrass, 1978, *Scientists Must Write* (London: Chapman & Hall), p. 107.
129	John Emsley, 'Lore and Order', *New Scientist*, 26 November 1988, p. 57.
139	Appears in Charles Taylor, 1988 *The Art and Science of Lecture Demonstration* (Bristol: Adam Hilger), pp. 1–2.
150	Quoted in Barbara Gastel, 1983, *Presenting Science to the Public* (Philadelphia: 151), p. 48.
151	Frank Albrighton, 1986, *Can I Quote You on That?* (Birmingham: University of Birmingham), p. 8.

Page no.	
152	Dorothy Nelkin, 1987, *Selling Science: How the Press Covers Science and Technology*, (New York: W. H. Freeman & Co.), p. 11.
153	Quoted in Sharon M. Friedman, Sharon Dunwoody and Carol L. Rogers, 1986, *Scientists and Journalists: Reporting Science as News* (New York: Free Press), p. 216.
158	Quoted in Sharon M. Friedman, Sharon Dunwoody and Carol L. Rogers, 1986, *Scientists and Journalists: Reporting Science as News* (New York: Free Press), p. 232.
161	Alec Nisbett, 1983, Speech to the British Association for the Advancement of Science annual meeting, University of Sussex.
167	From interview.

Index